PENGUIN BOOKS

THE BILINGUAL BRAIN

'A clear and approachable study by a renowned neurologist'
Miranda France, *Prospect*

'A fascinating primer on the science of language . . . Anyone with an
inquisitive mind and an interest in language in general, and certainly
bilingualism specifically, will find this book a compelling read'
Victoria Murphy, *TES*

'This lucid and fascinating book forensically explores how
different languages coexist within a single human brain'
Adrian Woolfson, *Wall Street Journal*

'Provocative . . . Costa speaks to the reader in a way that he and we
are equally helpless to resist – humanised, natural, charismatic'
Michael Hofmann, *London Review of Books*

'Neuropsychologist Albert Costa spent two decades exploring
bilingualism, and his book offers surprising insights'
BBC Science Focus

D1221781

ABOUT THE AUTHOR

Albert Costa was Research Professor at Pompeu Fabra University in Barcelona and director of the Speech Production and Bilingualism Group at the Centre for Brain and Cognition. His research focused on the cognitive and neural underpinnings of language processing.

ALBERT COSTA

The Bilingual Brain

*And What it Tells Us about
the Science of Language*

Translated by John W. Schwieter

PENGUIN BOOKS

PENGUIN BOOKS

UK | USA | Canada | Ireland | Australia
India | New Zealand | South Africa

Penguin Books is part of the Penguin Random House group of companies
whose addresses can be found at global.penguinrandomhouse.com.

First published in Spanish under the title *El Cerebro Bilingüe* by Debate 2017
First published in Great Britain by Allen Lane 2020
Published in Penguin Books 2021
003

Copyright © the Estate of Albert Costa, 2017
Translation copyright © John W. Schwieter, 2019

The moral rights of the author and translator have been asserted

Set in 9.12/12 pt Sabon LT Std
by Integra Software Services Pvt. Ltd, Pondicherry
Printed and bound in Great Britain by Clays Ltd, Elcograf S.p.A.

The authorised representative in the EEA is Penguin Random House Ireland,
Morrison Chambers, 32 Nassau Street, Dublin D02 YH68.

A CIP catalogue record for this book is available from the British Library

ISBN: 978-0-141-99038-5

www.greenpenguin.co.uk

This book is dedicated to my two favourite bilinguals: Chiqui, who makes the tuna croquettes, and Alex, who eats them.

Contents

Prologue

We are all talking heads. Even though we may not think about it, this is perhaps why everyone is consciously or unconsciously interested in language. From parents who watch their children with amazement as they say their first words, to people who suffer from speech problems as a result of brain damage, we have all wondered how the human brain acquires and processes languages. This book is dedicated to the age-old questions: how do two languages coexist in the same brain? And what are the implications of this coexistence? Is there anything special about being bilingual?

Why dedicate an entire book to the phenomenon of bilingualism? Well, because by and large, bilingualism is the rule rather than the exception in the sense that the majority of the world's population can communicate in more than one language. If we want to understand how language functions in the human brain, it would be a mistake to ignore this phenomenon. Studying the bilingual brain allows us to explore additional questions about how language interacts with other cognitive domains, such as attention, learning, emotion, and decision-making. Bilingualism is a window into the study of human cognition.

As you read this book, you might find more questions than answers about some of the issues raised. That is partly my aim: to pique your interest about how two languages coexist in the same brain and about how these issues are studied. During our journey, results from scientific research are presented to shed light on questions such as: how do babies exposed to two languages differentiate between them? What are the different language-learning trajectories for bilingual and monolingual babies? What are the neural bases that underpin the two languages of a bilingual speaker? How does bilingualism affect the development of other cognitive abilities? How do the two languages deteriorate due to brain damage? How does the use of a second language affect decision-making? If these questions seem a bit abstract, let's look at how it all works in practice.

Alex is a child born in a bilingual family in Boston where the mother speaks English and the father speaks Spanish. The parents each decide to speak to the child in their respective language, while wondering if this would have a negative effect on the baby's language development. They suspect that the cognitive processes that Alex would acquire for two linguistic systems would be different from those he would develop if they only spoke to him in one language (i.e., if the child were to grow up in a monolingual household). The parents know that their situation is not uncommon given that many babies are exposed to two languages, either because their family speaks two at home or because they have immigrated to a place where another language is spoken. Alex will have to learn to distinguish between the two languages in order to identify the sounds and words that correspond to each. In other words, he will have to develop two differentiated phonological and lexical systems. How is this achieved? Will this exposure result in a confused and deficient linguistic system? Well, it turns out that, despite what common sense or popular belief might dictate, we are beginning to have access to rigorous scientific evidence that explains the learning processes Alex will go through to acquire two languages simultaneously and without apparent difficulties. In this book, we will look at some studies that explore these issues as early as the first month of life. Yes, you read that right: babies who are only a month old. The ingenuity of researchers who are dedicated to studying human development is phenomenal. By the way, Alex is now fourteen years old and can get by perfectly in three languages: English, Spanish, and Catalan, and I should know because he is a chatterbox: Alex is my son.

Let's consider another case, that of Laura who was diagnosed with Alzheimer's three years ago and is still at an early stage of the disease. She lives alone in Barcelona, where she gets along quite well. Laura has always spoken to her daughter, Maria, in Spanish, as it is her first language, although she knows and has used Catalan regularly for the last eighty years. Maria begins to notice that her mother has some difficulties communicating and, while at the time it doesn't seem all that important, they begin to wonder about things like: when the disease progresses and more severely affects Laura's cognitive capabilities, what language will she end up speaking? Will the disease affect both languages equally? Will Laura be able to differentiate between the two languages and communicate in the one she wishes to without interference from the other? Studying these issues not only allows us, or better yet, *will* allow us to best help Laura (and her family), but it also informs us about how two languages are represented in a single brain. Knowing this will help us to make decisions about which language(s) to use in treatment and rehabilitation.

The cases of Alex and Laura are just two of the many examples that you will find throughout this book. Although I have taken the liberty of using pseudonyms to protect their identity, they are real cases that will help you to understand the issues that arise when we study how two languages are learned and used.

At this point you may be wondering what I mean by the term *bilingualism*. Although I prefer to avoid giving prescriptive definitions, I think it is fair to address this particular term. I promise that I will do it here and will not bring it up again.

Defining bilingualism is like trying to hit a moving target. By this I mean that traditional definitions are either so broad that they are not useful, or so narrow that they leave out many cases of people who use two languages. This is because the experience of being in contact with a second language is rather varied. For example, if we define bilinguals as only those people who have a very similar mastery of two languages, we ignore a large number of individuals who, although they perform much better in one language, use both frequently and without too much difficulty. On the other hand, if we take the age at which the two languages were learned as a defining characteristic of bilingualism, and consider bilinguals as individuals who have been exposed to two languages since 'the cradle', we will leave out another large set of people who use two languages regularly. To complicate the situation further, we often encounter great disparity in the abilities of the same person with respect to language use. For example, there are individuals who have advanced fluency and rich vocabulary in their second language but a very strong foreign accent. One such case is that of the well-known novelist Joseph Conrad, who wrote his most important works in English even though he was of Polish origin and learned English relatively late. Despite his extraordinary writing in English, Conrad had a strong Polish accent. You will probably agree with me that not considering him a bilingual would be absurd. By the way, Conrad is not an exceptional case; consider more recent cases (of perhaps less praiseworthy people) such as former US Secretary of State Henry Kissinger and former Governor of California Arnold Schwarzenegger.

It is obvious that we could go on dividing up, case by case, each of these groups of speakers and give them different names, although it would not be very useful because the number we would end up with would be too large. From my point of view, and with the evidence we currently have, it is more useful to treat each case as one among many points in a continuum of different variables (use, age of acquisition, competence, etc.), and not as differentiated groups. Of course, when conducting scientific studies, it is convenient that the sample of subjects be relatively homogeneous. In the following chapters, I will consider work done with different types of bilingualism, and although I will specify the characteristics of the different groups when necessary, I will continue to call them *bilinguals*.

We will start with the challenges that babies face when they are simultaneously learning two languages. In chapter 1 we will explore the different techniques developed to measure what babies know and do not know about each of their languages. Getting a baby who is only a few months old to give us meaningful data is not easy. I hope that you will enjoy discovering how researchers have developed ways to extract information from those little brains.

Chapter 2 is devoted to how the two languages are represented in the brains of adult bilinguals, paying special attention to studies from cognitive neuroscience and neuropsychology. We will see what areas of the brain are involved in the representation and control of two languages and how brain damage can affect both.

In Chapter 3 we will analyse the consequences that learning and using two languages have on language processing in general. Here, we will specifically focus on how the bilingual experience sculpts the brain by comparing brains of bilinguals and monolinguals. We will also review the extent to which bilingualism affects, both positively and negatively, language processing and why someone who is bilingual cannot be considered two monolinguals in one.

Chapter 4 will home in on how the bilingual experience affects the development of other cognitive abilities, particularly the attentional system. It is said, for example, that the continued use of two languages acts like mental gymnastics which results in a more efficient attentional system that is more resistant to brain damage. We will analyse to what extent the current evidence allows us to conclude whether this is the case. We will review studies conducted with individuals that cover a wide range of ages, from seven months to eighty years old, and recent work that suggests bilingualism can promote cognitive reserve in cases of neurodegenerative diseases.

This brings us to the last chapter, which shows how the use of a second language can affect decision-making processes. We will discover how the intuitive processes that sometimes skew our decisions are minimized when we use a second language. The studies discussed in this chapter explore economic and moral decision-making. Given that a large number of people participate continuously in negotiations in their second language (think about a multinational company or the European Parliament), the social implications of these studies are particularly important.

It is also worth mentioning two points that will not be covered in this book. Firstly, I will not be focusing on second-language learning methods. In other words, we will not discuss which strategies are most efficient for second-language learning in formal classroom settings. This does not mean that studies are not mentioned that have evaluated the effects of certain variables (e.g. age of acquisition) on second-language learning. However, this will be done in the appropriate context of each study, and not for the purpose of rigorously analysing which types of learning strategies are most effective for acquiring a second language in academic contexts. The second point that we will not address in this book are the social and political connotations that are often linked to the phenomenon of bilingualism, which have implications for educational models in many countries around the world. The coexistence of two languages in the same community and the discussions on identities that this often entails (think of the cases of the United States, Canada, or Belgium, to mention just a few) will not be discussed here.

At this point the only thing left to do is to invite you to join me on this journey to discover how two languages live in the same brain. Although at times during our exploration we will have to stop to discuss in detail a few supporting experimental studies, I hope that the journey will be engaging, entertaining, and informative. I also hope that the text will honour the Confucian proverb:

Tell me, I'll forget
Show me, I'll remember
Involve me, I'll understand

I

Bilingual Cradles

The Godfather: Part II tells the story of Vito Andolini, a boy around twelve years old who has fled by himself on a ship from his home town of Corleone, Sicily. As the ship arrives in New York City, Vito Andolini becomes Vito Corleone and so begins the Corleone saga in the United States.

Much like Vito, between the end of the nineteenth century and the first quarter of the twentieth century some 12 million people went through a United States immigration inspection on Ellis Island. Most of these immigrants, who were looking for a better future, came from European countries. When they arrived at the island, they had to answer a suitability questionnaire in which, among other things, they were asked about their country of origin, financial resources, and health. The luckier ones spent about five hours on the island and were eventually allowed to enter the country. Those who weren't so lucky spent much more time on the island, were placed in quarantine (where Vito was kept and suffered from smallpox), or were deported to their countries of origin. Interpreters, whose mission was to help newcomers formalize entry papers and interact with immigration officials, were key figures during this process. They were indispensable because Ellis Island was like a modern version of the Tower of Babel, where people came into contact with many different language backgrounds, from Italian and Armenian to Yiddish and Arabic.

These migratory waves were so intense it is estimated that currently around 100 million Americans share some kind of ancestry with the emigrants who passed through that island. Among them is my son Alex, whose great-grandparents also entered the United States there. It seems clear that a good number of these people prospered enough to form family ties for generations. It is difficult to imagine what it might be like to arrive in an unknown land with the intention of rebuilding a life far from one's country of origin. But it is relatively easy to imagine one of the challenges that many of these people had to face: learning a new language.

What does it actually mean to *learn* a language? It's not only about memorizing the words and understanding the grammar, but also acquiring the corresponding sounds (what we call *phonological properties*) and the appropriate use of expressions for a specific communicative context (which we call the *pragmatics* of a language). So it's not enough to know just the lexical labels (i.e. the words), but we must also learn the sounds of the language and how to combine them, learn which syntactic constructions are correct and which are not, and know what speech register we should use depending on our listener.

It is an enormous challenge to learn a foreign language and when we try to learn it as adults, in many cases it is only partially achievable. As adults, it is difficult for us to acquire sounds in a new language and so we have a foreign accent. It is challenging for us to acquire syntactic structures and, for this reason, in many cases we build sentences that contain grammatical errors, such as when, in English, someone says *we swimmed* instead of *we swam*. We may find it difficult to appreciate subtle aspects of meaning. This is why we sometimes use terms that are not appropriate for the communicative context, such as when we use swear words in situations in which they are not appropriate (try to explain to someone in which contexts it is appropriate to use the different swear words in your language). It is easy to get confused and see relationships between words from different languages to our own when there are none. For instance, some might think the word *embarazada* 'pregnant' in Spanish means the same as *embarrassed* in English. And, finally, as adults, it is difficult for us to coordinate all this information in a fluid way and therefore we get stuck when, despite our good intentions, we try to hold a conversation in the other language in question.

Yes, the challenge is enormous. However, things work differently when it comes to babies. We have all gone through this stage, and all of us learned a language with relative ease, or at least that's how it seems when we watch the linguistic development of children. How did we do it? In this chapter I won't be able to give an exhaustive answer to this question, nor even come close, but we will look at the challenges that babies face during language learning, in particular when it involves the simultaneous acquisition of two languages. The studies we will explore focus on the acquisition processes during the first months of a baby's development. I will discuss studies on monolingual babies and bilingual babies. Don't be surprised by my use of the term *bilingual baby*. While it is true that these babies do not *speak* either language yet, this does not mean that they don't have experience with both. The bilingual experience in many cases begins before babies are able to produce language and therefore the term is still appropriate, since it helps us to compare and contrast newborns who live with two different languages, and the challenges that this entails, with those who are only learning one. So we will call babies exposed mainly to a single language *monolingual babies*; and those who are systematically exposed to two languages *bilingual babies*. As we will see, there are some common challenges for both.

Before getting ahead of ourselves, it is important to remember that although babies do not speak, this does not mean that their brain is not continuously processing the information that is all around them. In fact, many studies have shown that during their first months of life babies acquire very sophisticated knowledge about language, and even if they won't start talking until after the first year of life (at the earliest), after a mere six months they will already have developed a complex knowledge of language, including a number of words.

WHERE ARE WORDS?

Let's start with a sentence in German from Goethe: '*Wer fremde Sprachen nicht kennt, weiß nichts von seiner eigenen.*' Those of us who do not know German won't understand any of it, but we will be able to guess that it contains ten words. It's easy: we will consider a *word* as every string of letters that has a blank space before and after (e.g. *Wer, fremde*, etc.). So, although we don't understand German, we have already taken a first step: we know that *Sprachen* is a German word, even if we do not know what it means. Now, put down the book for a moment and search for a song on YouTube in a language you do not know and listen to it carefully (if it's in German, maybe *Sprachen* appears and you will have already recognized a word). You can repeat the song if you'd like. Although you obviously will not understand what is being sung, are you able to identify separate words in the song's lyrics? In other words, are you able to guess where the white spaces are before and after the strings of letters? You probably have said no because you may have perceived the words as continuous strings of sounds without obvious boundaries between them. Don't give up yet; try again to cut the string of sounds into words. I bet that in many cases your word boundaries will not coincide with the real words or lexical items, and that you will group together sounds that belong to different words.

This shows us that, unlike written language, oral speech does not have well-defined 'blank spaces' between words and, if you had listened to Goethe's phrase above instead of reading it, you probably would have heard something like *WerfremdeSprachennichtkenntweißnichtsvonseinereigenen* and would have tried your best to figure out where one word starts and another one ends. I won't keep you in suspense any longer, here is what it means: 'He who does not know foreign languages knows nothing about his own.'

This is exactly what babies face when processing language. They experience situations in which segmenting speech into units that hypothetically can be words is somehow conducive to building their vocabulary, or mental lexicon. But how do they do it? And what happens when the same string of sounds can represent two words in two different languages?

Although it goes without saying, remember, all languages can be learned. If this were not the case, and there were a language that babies could not learn, it would disappear quickly. So, there must be some clue in the oral signal that allows babies to develop hypotheses about where to cut or segment the speech. That is, the string of sounds to which babies are exposed has certain regularities that should be able to guide their segmentation. For example, all languages have restrictions as to which sounds can cluster together. In Spanish, if we hear the sequence of three consonants *str*, there must be either a syllabic border between the *s* and *t* (e.g. *as-tro-nau-ta* 'astronaut') or a word boundary (e.g. *las trenes* 'the trains'). This is due to the fact that there are no words in Spanish ending with *-st* or beginning with *str-*. Surprisingly, by the time that babies learning Spanish are eight months old, they already realize this, even though they hardly know much language at all. How is this possible? One study which has had a significant impact showed that babies are able to compute probabilities of the co-occurrence of sounds. Let's pause for a moment to describe this study because it will help us to see how we can explore the linguistic knowledge that babies have from an early age.

In all human languages the probability that two syllables (or phonemes) follow one another (something known as *transitional probability*) is higher inside words than between words. For example, in Spanish, the probability that the syllable *pa* is followed by *la* (as in *palabras* 'words') is much higher than that of the syllable *bras* followed by the syllable *que* (as in the phrase *las palabras que oímos* 'the words that we hear'). Jennifer Saffran and colleagues at the University of Rochester carried out a study testing the hypothesis that eight-month-old babies are capable of computing transitional probabilities between sounds. The researchers created sequences of syllables in which the transitional probabilities between different syllables was manipulated (see Figure 1.1). So that there was not an effect of the baby's own language, in this case English, the experimental words were invented in a way such that they had nothing in common with English. The trick was that there were sequences of syllables that formed what the researchers deemed as *words*. The transitional probability between the syllables of the words was 1 (i.e. 100 per cent). For example, one of the words was the sequence *tupiro*. Thus, whenever the syllable *tu* appeared, the syllable *pi* followed, and whenever the syllable *pi* appeared, the syllable *ro* would follow. After the sequence *tupiro*, another word from the experiment appeared (e.g. *golabu*, *bidaku*, *padoti*) as long as its probability of appearing after *tupiro* was 0.3 (i.e. 30 per cent) (e.g. after *ro*, the syllables *go*, *bi*, or *pa* could appear).

The probability of transition between the syllables of different invented words was much lower (only occurring one-third of the time) than the probability of transition of syllables within words, which occurred in a deterministic way. To put it another way, there

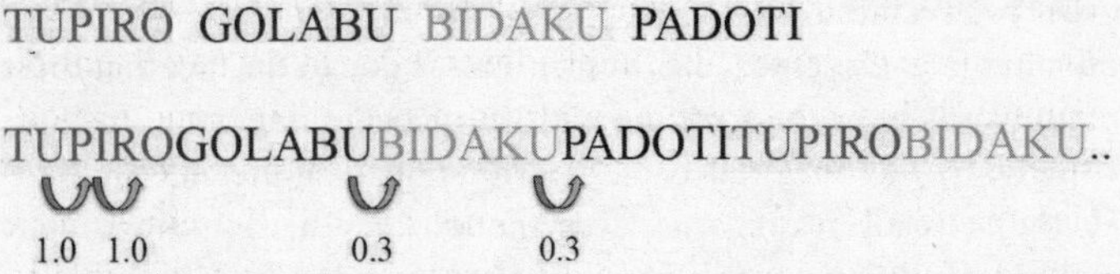

TUPIRO GOLABU BIDAKU PADOTI

TUPIROGOLABUBIDAKUPADOTITUPIROBIDAKU..

1.0 1.0   0.3   0.3

Figure 1.1. A transcription of the order in which syllables were presented in the experiment. As can be seen, the syllables of each invented word always follow the same order. For example, whenever *pi* appears, *ro* will always follow. However, when finding *ro*, the next syllable can be *go*, *bi*, or *pa*.

were syllables that tended to appear together very frequently, and others much less often. This experiment sought to simulate the situation that we described above in the example of *las palabras que oímos*: there are some sequences of syllables (e.g. *pa* and *la*) that tend to appear together more often than others (e.g. *bras* and *que*). The researchers played the strings of syllables to the babies for two minutes, without intonation or pauses between the syllables. In fact, the recordings were reproduced by an artificial sound-generation system which produced sounds a little worse than what you may have heard when you listened to a song in an unknown language.

Would eight-month-old babies be able to compute these regularities, extract patterns from strings of sounds, and learn that certain syllables always go together, and others do so less often? If this were the case and babies were able to compute probabilities, and specifically transitional probabilities between syllables, then it is possible they would realize that the sequence *tupiro* always appeared together (i.e. it formed a word on its own) and that the sequence *rogola* appeared less often and never as its own word. This would suggest that babies are able to pick up on the statistical regularities present in speech as a segmentation strategy to detect words.*

This all sounds straightforward and I hope that you agree with

---

* This is only one possible strategy, which alone is perhaps not sufficient to account for speech segmentation processes. In fact, we know that speech contains other clues that babies rely on, for example, the alternation between stressed and unstressed syllables or syllable duration.

me that the study is as elegant as it is simple. But how do we ask eight-month-old babies? We simply observe how they pay attention to stimuli that make up words and those that correspond to non-words after playing the string of syllables for two minutes. If the babies reacted equally to both types of stimuli, the experiment would fail and we would have no indication that they are able to identify syllable strings that appear together more often (and if this were so, I would probably not be telling you about the study). But the babies *did* pay more attention to stimuli that, during the familiarization/training stage, did not form a word compared to stimuli that did. We know this because, when hearing these stimuli, the babies spent more time looking at the sound source and were less distracted. It was as if they were surprised by those stimuli that, despite them having been heard during the training stage, had not been segmented as words. The origin of such a surprise rested in the fact that the babies had acted as statistical machines during the training phase, unconsciously computing transitional properties between the monotonic strings of sounds to which they had been exposed.

So what must have occurred in babies' heads was something like: 'If the sound *tu* appears, it is very likely that *pi* and *ro* will follow, and as such, it must be a unit of some type . . . a word; whereas if *ro* appears, it is unlikely that *go* follows, and that's why the sequence *rogola* does not appear to be a word.' And you thought that babies only eat, sleep, and . . . well, think again! The next time you see a baby, remember that there is a powerful statistical computer in front of you.

We stopped to describe this study in detail because it is an example of the types of experiments that are carried out to learn more about which phonological clues babies use to make sense of the strings of sound to which they are exposed. Thanks to studies like this one, we now know that babies are sensitive to many of the regularities present in the speech signal, such as the possible combinations of sounds in a language (what we call *phonotactic rules*), the patterns of intonation and accentuation, and the repertoire of sounds. Although the sensitivity to each of these properties varies according to age, they all help babies to extract the words from speech that allow them to build a *lexicon*, a mental dictionary.

WHY ARE THEY DOING THIS TO ME . . . AND WHAT IF THE TWO LANGUAGES DON'T AGREE?

As if the task of decoding speech signals during the first months of learning weren't already hard enough, babies who are exposed to two languages at the same time must face additional challenges. Although, as we have seen, there are certain phonological regularities in all languages, these do not necessarily have to be always the same across languages, and, in fact, they are not. Returning to our example of the sequences of sounds that are permissible in a language: in Spanish, there are no words that start with *str-*; thus, a baby with sufficient exposure to this language may tend to consider that the sequence *risas tristes* 'sad laughter' contains at least two words and that there probably is a boundary between the *s* and the *t*, at least a syllabic boundary. English, on the other hand, contains many words that start with the sequence *str-* (e.g. 'strong', 'stream', 'strange', etc.) and, therefore, a baby exposed to English should not have the tendency to identify those two sounds as belonging to different syllables or words. For this baby, to assume that the sequence *four streets* has a word boundary between the *s* and the *t* would be counterproductive, since it would result in *fours treets*. And now comes the hard part: how will this situation be handled by a baby exposed to both Spanish and English? Using either one of the two strategies, whichever one it may be, will have negative repercussions when trying to apply it to the other language. The confusion can be tremendous, but in reality, there does not seem to be a significant delay in extracting statistical regularities even when babies are exposed to two languages.

On the other hand, there are phonological properties that are relevant to one language but not to the other. For example, there are *tonal languages*, such as Mandarin Chinese and Vietnamese, in which the same syllable can take on different tones to correspond to different things. That is, if we enunciate a syllable with a higher or lower tone, we will be referring to different meanings. This is what we call a *contrastive property*: tone (the fundamental frequency with which we produce a sound) is used in differentiating lexical items. In Spanish, the intensity with which a syllable is emitted within a word also corresponds to a contrastive property. It's what we call *stress*. There are words that differ only by stressing one syllable over another as in the pair of words *sábana* 'sheet' and *sabana* 'savannah'. Later we will return to talk more about phonological contrasts.

Tone in Mandarin would work equally in terms of its contrastive value. This language has at least five tones. So for example, the syllable *ma* can mean five different things depending on the tone used: *mā* 'mother', *má* 'to numb', *mǎ* 'horse', *mà* 'to scold', and *ma*, an interrogative particle. In fact, you can create a *raokouling*, or tongue twister, with *ma*: *māma qí mǎ, mǎ màn, māma mà mǎ,* which means 'mother rides a horse, the horse is slow, mother scolds the horse'. By the way, if you are wondering whether this tonal property is a rarity among the world's languages, then you would be very surprised to know that actually 40 per cent of languages are tonal. However, this property is not contrastive in Catalan or English, or in many other Indo-European languages. Although the syllables can be pronounced with different tones, this is irrelevant from a lexical point of view. In Spanish, the sequence *pan* 'bread' has a meaning regardless of the tone in which it is said. So, although a baby exposed to Mandarin Chinese will have to learn to be sensitive to the tone with which the syllables are used, a baby learning Spanish will have to learn to ignore that property, at least as being a contrastive lexical property. This is yet another challenge to overcome for bilingual babies.

It is clear then that a baby exposed to two languages from the cradle has to learn that certain clues in the speech signal are relevant for one language and not for the other. But for this to happen, they will have to first realize that there are two languages at play. In other words, they will have to realize that they are in a bilingual environment. And when they do, they might wonder, 'Why are they doing this to me? Wasn't it difficult enough already?'

AHA! MY PARENTS DON'T
SOUND THE SAME

At the end of the day, babies exposed to two languages end up learning both with no problem, so, although they might complain (if they do at all), they end up resolving the situation well enough to become bilingual. What is more interesting to find out is how and when a baby is capable of realizing that the strange sounds that come out of one parent's mouth have different properties than those that come out of the other parent's. In other words, do babies realize that there are two linguistic systems going on around them? Before getting to the answer, let me digress a little to show you how we are born already sensitized to speech signals.

In a study carried out in Trieste, Marcela Peña and her collaborators studied the brain activity of newborns when exposed to a speech signal. More specifically, the researchers wanted to know to what extent there was a preference for language processing in the newborns' left hemisphere as is commonly seen in adults. To do this, they measured the brain activity of sleeping two- to five-day-old babies while being exposed to two different types of stimuli. First, normal language was played using stories that were read by mothers of babies who did not participate in the study. Second, those same stories were reproduced backwards (i.e. the speech was played back from end to beginning). Obviously, the speech in the second stimuli shared many acoustic properties with the normal speech (e.g. the volume itself), but it was immediately apparent that it was not a language. If you are old enough, you may remember the sound that a cassette tape made when pressing on the play and rewind buttons at the same time. Well, the second stimuli sounded something like that.

Would a two-day-old brain be able to differentiate between the two speech samples? The answer is yes. When the normal language version was presented, the brain activity of the babies, measured by oxygen consumption in the brain, was greater than when the story was reproduced backwards. Moreover, the difference in brain activity for the two stimuli was mostly present in the left hemisphere, the one that is most involved in language processing in general. Thus, the brain of the newborns not only reacted differently to the two versions of language samples, but it was precisely the hemisphere involved mostly in language that selectively responded. These findings indicate that our brain is born skewed to interpret language in a certain way. The ability to differentiate between these two types of stimulus (normal and backwards speech) does not imply that newborns are able to differentiate between two languages.

A sizable number of studies have shown that babies are able to discriminate between languages that sound quite different, as soon as hours after birth. Yes, *that* soon after birth. Furthermore, this ability does not require that babies be exposed prenatally to such languages. A newborn of a mother who speaks Spanish will be able to discriminate between, for example, Turkish and Japanese. Obviously, the baby will not know what he is hearing is Turkish and Japanese, but he will know they are different or, rather, that the sounds are different. And if this ability does not surprise you, try being unmoved by this: some types of monkeys and even rats are also capable of differentiating languages that have very different phonological properties. This suggests that certain human abilities that have to do with language processing are also present in other species that do not end up developing a language as sophisticated as ours.

It is worth pausing for a moment to ask how it is possible to know whether newborns are capable of differentiating between two languages. How do we ask them this question? The answer in this and many other cases has to do with the effects that familiarity with various stimuli cause in babies. When we subject them repeatedly to stimuli of a certain type (until they are bored, in fact) and in a later phase they are shown the same or different stimuli, their behaviour changes depending on the stimulus presented. In general, they show a greater preference for novelties (things that have not been shown before). That is, there is a difference in behaviour when a new and an old object are presented, so we can learn about what information babies are processing. We can explore this preference by measuring the time they spend paying attention to a certain stimulus: the more novel it is, the longer they will fixate their gaze on it. *New is cool!*

Studies with babies who have been born within a few hours employ the non-nutritive sucking method. The method works based on the fact that babies have a sucking reflex from birth which also corresponds to their attention level. The more attention, the more sucking. If we expose a baby to a repetitive stimulus, for example, the chain of syllables *ba, ba, ba, ba, ba*, we will see that its sucking rate and/or strength decreases as the stimuli are presented (or rather, repeated). Sucking rate is measured with an electronic pacifier that records each reflex movement the baby makes. Don't worry, it's not an invasive method at all, it's just a normal pacifier that has an electronic sensor allowing us to measure properties of each suck. What the baby is possibly telling us with the decline in sucking is 'I've caught on, you keep presenting me the same thing and I'm getting bored of hearing it continuously.' If that were so, a change of stimulus would make the baby quit being bored and consequently increase the sucking rate; remember: *New is cool!* Of course, that would happen if the baby were able to notice the difference between the stimulus that is boring him and the new one; if he were not able to pick up on the difference, he would continue being bored. Following our example above, the sucking rate increases when the sequence *pa* is inserted into the string *ba, ba, ba, ba, ba*. This implies that the baby has noticed a change, a difference between what was repeated and ended up boring him (*ba*) and the new stimulus (*pa*). In fact, as we shall see below, this technique has shown us that, shortly after birth, babies are able to discriminate between the sounds of all languages.

Now that we have a way of 'asking' babies what things they can and cannot differentiate, let's return to the question at hand: can babies distinguish between two languages like, for example, Turkish and Japanese? To find out, babies are exposed to a series of sentences in one of the languages (e.g. Turkish). At a critical point, either a sentence in the same language or in the other language (let's say Japanese) is presented. If the sucking is different between the two languages, then it's clear the baby can differentiate between the languages.

This type of research, of which my mentor Jacques Mehler was a pioneer, has shown us what types of languages can be distinguished at very early ages and what types cannot. It has also helped to group languages into different families depending on their phonological properties, that is, how they sound. Again, it is possible that not all sound similarities between different languages are equally important to the baby. What interests us is finding out to what properties babies pay attention when differentiating languages, because in some way they are giving us information about what phonological properties are most important when it comes to learning them. From research, we now know that the ability to differentiate between languages from different phonological families appears very soon. For example, differentiating a string of Dutch sounds from Japanese is relatively easy. However, the ability to differentiate sounds in two languages of the same phonological family appears a little later, and in fact, it is necessary that at least one of them be known to the baby. For example, an Italian baby will be able to differentiate Spanish from Italian, but it will be much more difficult for that baby to do so between Spanish and Catalan, even though all three are Romance languages. Exposure to a language, therefore, is fundamental in separating it from similar ones.

The fact that newborns can differentiate between two languages does not guarantee per se that bilingual babies do not experience some degree of confusion. It's one thing for the baby to distinguish between dissimilar languages to which he is not usually exposed, and another to have a certain degree of confusion when being exposed to them. Moreover, one might think that such confusion would be greater if the two languages were from the same phonological family – the more two things seem alike, the more they seem to be the same thing. The question is, to what extent does bilingual exposure mean the capacity to discriminate between languages is viewed as favourable or as an interference?

Although the information we have about this question is somewhat limited, we know, thanks to the studies by Núria Sebastián and colleagues, that at four months old, Spanish-Catalan babies are already able to differentiate between languages that are as similar as these two. In fact, monolingual babies who only understand Spanish are also capable of doing it. However, it seems that bilingual and monolingual babies do it differently. Monolinguals are quicker to look at a source of sound when it corresponds to their first language than when it corresponds to a language not known to them. Let me explain. The study in question measured the time it took for the babies to orient themselves to a sound source when it corresponded to one or another language (see Figure 1.2). To do so, the babies were shown a visual stimulus on a computer screen. When the baby looked at the stimulus intently, that is, he fixed his gaze on it for a few seconds, a phrase was heard from a speaker on the side of the screen. This speaker was covered with the drawing of a woman's face. The phrase could be in the baby's first language or in an unknown language. The results showed that monolingual babies tended to look at the source of sound more quickly when the phrase corresponded to the first language than when it corresponded to the unknown language. For the bilingual babies, the opposite happened. We still do not have a convincing explanation for this phenomenon, although it may be that bilingual babies are evaluating which of the two languages is the one being heard, something that would take additional time. But this is just a hypothesis. What matters for our purposes here is knowing that bilingual babies are able to differentiate their two languages from other languages.

The last thing I'd like to mention about differentiating languages is the effect of the bilingual experience before birth. Yes, you read correctly, *before* being born. In principle, it might not be the same to be in the womb of a mother who only speaks one language as being in the womb of one who speaks two languages. We know that at birth babies can already differentiate the mother's voice from other voices. In fact, they show a preference for words spoken by their mother compared to those uttered by a stranger. This is not very surprising,

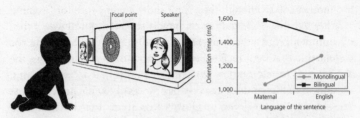

Figure 1.2. An example of what the experiment might look like (left). The graph (right) shows the orientation time of monolingual and bilingual babies towards their first language or English. As can be seen, monolingual babies are more quickly oriented to their first language whereas the bilinguals are more quickly oriented to the foreign language.

since, after all, they have been hearing that voice for around nine months and although it's true that the sound conditions in the womb aren't the best, they must be hearing something. From a survival point of view, it is obviously beneficial to know whether the person talking is your mother or a stranger. However, newborns show a preference not only for their mother's voice but also for the language that she used during the pregnancy. Two-day-old babies whose mothers have spoken Spanish during the pregnancy tend to prefer to listen to phrases in Spanish even if spoken by a stranger, and those whose mothers speak English tend to prefer hearing phrases in English. Apparently, nine months in the womb go a long way.

If it is true that babies are 'exposed' to their first language before they are born, the question is, what happens if this experience involves two languages? Will they think they have two mothers? Will they mix up the words as if what they had been hearing in the womb belongs to the same language? This last option would be logical, given that those sounds that they hear come from the same person. But it turns out that neither option is correct. One study showed that babies whose mothers used Tagalog (the majority language of the Philippines) and English when they were pregnant did not show a preference for either language at birth. Does that mean the babies are not able to tell the languages apart? Or that they are mixing them up? After all, babies who have listened only to English during pregnancy show a preference for that same language. It turns out that they are not mixing up the languages, nor do they think they have two mothers. Even though they don't show a preference for listening to either Tagalog or English at birth (why should they anyway if they are both spoken by their mother?), when they are tested using the non-nutritive sucking technique, they show evidence of noting the difference between the two languages. In other words, the prenatal bilingual experience does not confuse the baby, so it's perfectly fine to speak the languages you want when you are pregnant. There is no problem, although I think you already suspected that.

VISUAL CUES AND BUILDING A SOUND INVENTORY

The fundamental medium used by language is sound. We spend much of the day talking and, once in a while, listening. This is also how things go after learning to read and write, or even when communicating through social media. Remember, we are *talking heads* and not so much *writing heads*! Or, as Charles Darwin elegantly put it in *The Descent of Man* (1871): 'man has an instinctive tendency to speak, as we see in the babble of our young children; whilst no child has an instinctive tendency to brew, bake, or write.'

While we speak, the auditory signal is generally accompanied by other types of clues that affect our perception of it. When we talk to someone (if it is not by phone, of course) we tend to watch the movements that the other person makes with their mouth, looking at the lips and the articulatory movements that produce the sounds. This becomes more evident in situations where listening comprehension is difficult, such as when there is background noise (e.g. when we are in a bar or concert) or even when we interact in a second language that it is hard to understand. I am sure that you have noticed this tendency when, for example, watching a movie when the audio and the image on the screen are not in sync. Just the slightest mismatch between the two makes for an irritating experience. This shows that human beings gather visual and auditory information almost automatically while talking to someone. (If you want to have some fun, you can experience an audiovisual illusion called the McGurk Effect, which perfectly exemplifies this phenomenon. Just search for 'McGurk Effect' on YouTube and watch a video.)

But what does this have to do with bilingual babies? It turns out that babies seem also to use visual cues to discriminate between languages. Babies between four and six months old are able to differentiate between French and English only by watching videos of people speaking in those languages without sound! This capacity is maintained until eight months for babies who have been exposed to two languages, but not for those who have been exposed to only one. Bilingual exposure seems to strengthen and lengthen the ability to focus on the articulatory movements of the lips in order to differentiate languages.

In fact, it seems that there is a very early bias associated with the bilingual experience towards paying attention to articulatory movements. Four-month-old bilingual babies fix their gaze on the mouth of a person who is speaking to them longer than monolingual babies. This bias is maintained for at least a year and suggests that the complexity of being exposed to two languages encourages babies to extract as much information from communicative acts as possible (whether they are acoustic or visual information) in order to differentiate between languages. And you thought that they only ate, slept, and . . . but no, they are all but ready to jump!

The fact that bilingual babies can discriminate between languages only with visual cues may not surprise you too much . . . although I challenge *you* to try it. Turn off the volume of the television and try to guess in which language the actors in a movie are speaking. Good luck. But the effect of the bilingual experience on the ability to read the lips goes further. It turns out that at eight months bilingual Spanish-Catalan babies are able to discriminate visually between two languages to which they have never been exposed (French and English), whereas monolingual babies (either Spanish or Catalan) are not. There definitely seems to be something happening with these bilingual babies who like to watch lips so much.

We began this chapter describing a study that explored babies' capacity to extract the statistical regularities present among the different sounds of a language. We saw how that skill is very useful in segmenting language and discovering the permissible sequences of sounds that are candidates to form words. Similar to this skill development, babies must learn the distinctive sounds that are present in language, known as *phonemes*. Phonemes are the smallest units that differentiate two words. For example, the words *bat*, *rat*, *cat*, and *mat*, are all different just by the first phoneme. The difference between, for example, the *b* and the *r* is called *contrastive* since the alternation of one phoneme or another gives us different words, the same as what happened with tone in Mandarin Chinese. Having difficulties acquiring contrastive phonemes can lead to mistakes such as saying *lice* instead of *rice*. Have you heard this error before? We will talk more about it later.

One of the primary jobs for a baby during the first months of learning is to build what we call a *sound inventory* of the language to which he or she is exposed. Put another way: the baby has to learn all the sounds that have been coming out of his or her parents' mouths.

Babies are born with the capacity to acquire any phonological properties that are present in human languages. If this were not so, as we have argued above, those languages that were not acquirable would disappear. So, although it seems incredible to us, six-month-old babies, for example, are able to discriminate between sounds belonging to languages to which they have never been exposed with almost the same precision as babies who *have* been exposed to them. The problem that some adults whose first language is Chinese have with *l* and *r* is not present in babies who are born in environments where the language is spoken.

But this ability to recognize phonological contrasts is reduced as babies grow older. In a study that has become a classic in first-language acquisition, carried out by Janet Werker and her collaborators, the ability was evaluated for babies of different ages to discriminate between sounds of their first language compared to sounds of a language to which they hadn't been exposed. In the study, the ability to differentiate two Hindi phonemes, which are not contrastive in English, was explored among six- to twelve-month-old babies who were growing up in either English- or Hindi-speaking environments. That is, these two sounds serve to distinguish between words in Hindi but not in English. Therefore, for those babies who are learning Hindi, the difference between the two sounds is important because they can give rise to different words, whereas such a difference is irrelevant for babies exposed only to English. In fact, babies learning English would do well to ignore this insignificant difference. The study showed that babies at six months were able to discriminate between the two sounds regardless of what language they were learning. However, by twelve months old, only babies who had been exposed to Hindi were able to discriminate between the two Hindi phonemes. That is, after just twelve months of exposure to a language in which the contrast between the two sounds in question was not relevant, the ability to differentiate those sounds was lost (or at least significantly reduced). This shows that the passage of time is critical in terms of our ability to distinguish sounds.

This result and other similar findings are important for several reasons. First, they show us that at least during the first six months or so, babies are sensitive to sound contrasts that they have never heard before. Second, this ability seems to be lost at an early age if there has not been natural exposure to the phonological contrast in question. In addition, this loss of sensitivity is accompanied by an increase in sensitivity to detect subtle differences between the phonemes of the language to which the baby *is* exposed. This phenomenon is called *perceptual narrowing* and, as we have just seen, it has two implications. As we continue to learn a language, our ability to process its phonological elements increases while our ability to process phonological elements in another new language decreases. At first glance, this perceptual adaptation would seem to be a disadvantage but, in fact, it is a great adaptation feature, because it allows us to 'separate the wheat from the chaff', that is, it allows us to pay attention to the relevant information in the environment (the wheat) and, in turn, to ignore things that are irrelevant (the chaff). However, this perceptual narrowing comes at a price, since decrease in the ability to process phonological elements of a language to which one has not been exposed from very early on could explain a foreign accent in a second language.

You may think that having to process two different languages would significantly affect perceptual narrowing. That is, it could be that the increase in the variability which bilingual babies face in two languages reduces the awareness of the relevant information of each of them. However, this phenomenon is present in both bilingual and monolingual babies around the same age, so it does not seem that exposure to two languages alters development at this level of linguistic processing.

In fact, the bilingual experience does not seem to alter the establishment of phonological categories either. For example, in one study, at twelve months, bilingual French-English babies were able to discriminate between contrasts that appeared in their two languages in the same way that monolingual English babies were already able to do in their language. The truth is that bilingual babies are like machines.

One of the challenges that bilingual babies face while developing their sound inventory is that sometimes two phonemes belonging to each of their languages can sound very similar but not exactly the same. For example, consider the sound /b/ in English and Spanish. At first glance, they appear the same, but looks can be deceiving. The Spanish /b/ tends to have some voicing before we open our mouths whereas the English /b/ does not. By voicing, I am referring to the fact that the vocal cords vibrate a little before we open our mouths to articulate the sound. Try it for yourself: put your hand on your throat and try to produce the syllable /pa/ but without emitting the sound /a/. You will see that the vocal cords do not vibrate, or at least they do not vibrate until you open your lips to say the vowel. Now, try to pronounce the syllable /ba/. Before releasing the lips, you will notice how the vocal cords vibrate a bit. That is the difference between /ba/ and /pa/: the first one is voiced before opening the mouth and the second one is not, a very fine detail that happens in milliseconds.

Learning the difference between how to pronounce the English /b/ and Spanish /b/ is very difficult, and is partly the reason for a foreign accent. How do bilingual babies overcome this situation? We know that at ten months old they are able to distinguish between the two /b/s. In contrast, the monolingual cannot; all of the /b/s sound the same, probably like the /b/ in his language. The monolingual baby has grouped all those variations of the /b/ under the same category whereas the bilingual has separated them into two categories, the English /b/ and the Spanish /b/. It is clear, then, that bilingual babies are able to structure a differentiated sound inventory for each of the languages.

We know that babies of around nine months old have had sufficient experience with language(s) to show a certain amount of sensitivity to phonotactic rules. We believe this to be true because babies demonstrate a preference for hearing words that contain highly frequent sound sequences in their language rather than less frequent sequences. In the case of bilingual babies we are faced with the same situation as before: not only are they having to do twice as much work, but they must also ensure that the statistical information computed for one language does not mix with the other.

Are babies able to build these regularities separately? The information available to us on this issue is very limited. In a study carried out in Barcelona, researchers explored to what extent monolingual babies of Spanish or Catalan show preferences for strings of sounds that are phonotactically possible in Catalan (i.e. they often appear in that language) compared to strings that never appear. The results were clear: at ten months old, babies exposed only to Catalan showed a preference for permissible sound sequences in that language, whereas babies exposed only to Spanish did not. Up to here, everything makes sense: if you expose babies to the input, they will get it; if not, then they won't. But what about bilingual babies? It turned out that bilingual babies whose first language was Catalan and who had greater exposure to it, performed the task like the babies monolingual in Catalan. Babies from bilingual environments, but with Spanish being more dominant, did not show such a marked preference for the sequences of Catalan sounds. This difference may be due to the variation in the quantity of linguistic input that the baby receives in the non-dominant language, which seems to affect the development of the phonotactic rules of both languages, at least at this very early age.

WHAT DO WORDS MEAN?

Up until this point we have discussed some of the strategies that babies use to segment speech signals and find sequences of sounds that could be candidates to form words. The truth is that I've done a bit of cheating, but only a little. It is one thing for babies to be able to *recognize* strings of sounds that tend to go together, but another thing for them to *discover words*, in the sense of how *word* is commonly defined. In the German phrase by Goethe I quoted earlier we could detect which sequences of letters went together, but we did not know what they meant; therefore, we were not discovering the words. This implies associating the string of sounds in question with a reference in the real world, be it an object, an idea, or a property. So, when we learn a language, we not only have to realize that the sequence *dog* appears frequently, but also discover that this sequence refers to the domestic animal that we like so much.

As we have seen in previous sections, for their first six months, babies are relatively good at detecting patterns of sounds that frequently appear in a speech signal (remember the string of sounds *tupiro*?). And, in fact, at this age they are already able to associate some of these chains of sounds with their referents, as long as they are highly frequent. As parents well know, there is a very significant increase in word learning around a year and a half. At this age, children begin to discover words at a more constant and faster rate (averaging about ten per week). This is often referred to as the 'word-spurt period'. However, these observations refer to the vocabulary inventory that the children are capable of *producing* at that age, which does not imply that at younger ages they have not already learned many of those words and were only able to recognize them when *hearing* them. That is, it is one thing for children to be able to say the words aloud, but another thing for them to understand the words when hearing them.

A study conducted by Janet Werker and her collaborators at the University of British Columbia shows how researchers explore these abilities in babies. The researchers wanted to find out at what age babies were able to associate words with their referents. To do this, babies were taught two new words for two unknown objects. In the first training session, the babies saw a new object (object 1) while hearing a new word (word 1) and in the second training session, another new object (object 2) was presented with another new word (word 2). Each duck was in its line, so to speak. The question was whether the baby would realize this. If this were true, when the objects were mismatched with the words, that is, when a duck fell out of line, the babies would be surprised. That is exactly what happened in the study: after the training sessions, the babies were presented with each object along with its corresponding word or with the other word that didn't correspond. Thus, object 1 could appear either with word 1 (consistent pairing), or word 2 (inconsistent pairing), and object 2 could appear either with word 2 (consistent pairing), or word 1 (inconsistent pairing). The results were clear: the babies gazed longer at objects that appeared with an inconsistent pair than when they appeared with a consistent pair. In other words, the babies were surprised by the fact that these objects appeared with a verbal label that did not correspond to the one they had just learned. Simple and elegant, don't you think? Interestingly, this discrimination only occurred after the age of one, not before.

Having come to this point, we should ask to what extent the

capacity to associate objects with their referents is seen in bilingual babies. There are reasons to believe that the learning mechanisms for bilingual babies could be relatively different from those for monolingual babies. And, by definition, for bilingual babies, any object can carry two associated words: a word in one language and a word in another language. Bilingual babies face, then, a more variable linguistic stimulus. Researchers who have studied the association between objects and new words have observed that at fourteen months old both bilingual and monolingual babies were surprised when the association between objects and words established during the training phase did not match. In addition, both groups showed no such surprise at twelve months, suggesting a common development trajectory for both. Therefore, it does not seem that the bilingual experience affects the ability to relate objects and words.

However, there do seem to be certain differences in some of the strategies that come into play when bilinguals and monolinguals learn new words. One of them is what we call the *mutual exclusivity strategy*. This strategy is based on a bias that all children and adults possess: that every referent or object of the real world has only one word to describe it. This helps babies to hypothesize that if someone refers to a known object using a different word than the one they already know, then the referent must correspond to something else, like some part, substance, or property of the object in question. This mutual exclusivity strategy leads the child to develop a disambiguation bias, which is very useful for linking new words to new references. For example, if we show an eighteen-month-old child two objects, one that we know he knows (e.g. a stuffed bunny) and one that does not exist in the real world (e.g. some sort of rhino-frog hybrid), when we say a new word, the child will tend to look at the object he does not know (the 'rhinosofrog') and not so much at the other. It is as if the child realizes, 'If I hear a new word, and I know that the furry, big-eared object in front of me is called a rabbit, then it is likely that the new word refers to the other object that I've never seen.' This seems to be a good strategy for building vocabulary.

Again, the case of bilinguals is a bit more complicated, because for them, objects can carry two associated words. That is, the new word could also refer to a stuffed bunny, but in the other language.

Consequently, using the principle of mutual exclusivity for the bilingual baby could be risky. In fact, some studies have shown that the disambiguation bias is less prevalent among bilingual and, especially, multilingual children. That is, bilinguals do not tend to look more at the new object when a different word is presented. In addition, the application of the disambiguation bias seems to depend on how many translations the baby already knows, as reported by the parents. Children who know more translations in their two languages tend to show less application of the disambiguation bias. Therefore, it would seem that the bilingual experience, which involves the learning of translations that refer to the same object, reduces the use of the mutual exclusivity strategy. It makes sense.

What is still left for us to find out is what other strategies bilingual babies use to help them compensate for the reduction of use of the mutual exclusivity strategy. I say *compensate* because, when we count how many words a baby knows, it turns out that bilingual babies know more than monolingual babies. Let me clarify this a little. The number of lexical labels that bilingual babies recognize in each of their languages is lower than that of monolingual babies. However, when we consider their total number of words, that is, when adding up the ones they know in both languages, the number is greater than that of monolingual babies. Furthermore, knowing fewer words in each language does not mean that bilingual children have fewer concepts associated with words, that is, knowing that the furry animal that barks corresponds to a word, whether it be *dog* or *chien*. In fact, bilingual and monolingual children do not differ in this respect, and manage to associate concepts with words (but not always in both languages at the same time) at a similar rate. So, one should not worry if bilingual babies have a delay in word acquisition; they simply have twice as many to learn. We will return to this question in Chapter 3.

SOCIAL CONTACT IN LEARNING

When I was in high school, there was a saying that if you slept while playing a recording of a lesson you had to learn, the information would be learned while you slept. I suppose that the same people who came up with this learning strategy also believed that it worked for foreign-language learning. I must confess, I tried it once when I was learning English in high school, and given the grades I received in my English classes, it seems that it did not work out too well for me. I tell you this because there are many parents who want their babies to learn a second language from an extremely early age. Sometimes this leads to the belief that simply exposing the baby to a second language can lead to acquisition. After all, if babies are so good at extracting statistical regularities from speech signals with what appears to be almost automatic efficiency, it seems reasonable to think that merely exposing them to a new speech signal would be sufficient for them to learn it. Unfortunately, while it might sound conceivable, it is likely impossible. The mere passive exposure to a language is not very effective. In fact, social interaction is fundamental for language acquisition, even for the learning of representations as basic as sounds. Let's take a look at an example.

A study conducted by Patricia Kuhl at the University of Washington explored the acquisition of sounds belonging to a foreign language. Kuhl designed a learning protocol in which two groups of nine-month-old babies monolingual in English interacted with an adult tutor while playing or reading books. One group had a tutor who spoke to them in Mandarin Chinese (a language they had never heard before), and the other group (the control group) had a tutor who spoke to them in English. After this learning period, or interaction period, if you will, which lasted about four weeks, the researchers analysed the babies' ability to discriminate between a phonological contrast that occurs in Mandarin but not in English. Would children exposed to Mandarin have learned that contrast? Yes, they were able to discriminate between that contrast not only better than the children in the control group, but even with the same success as children exposed to Mandarin for ten months.

These results are very interesting because they demonstrate the speed and reliability with which children can acquire new sounds (or at least improve their learning). This is good news. But if it is so easy to learn a phonological contrast, maybe a tutor is not necessary. Perhaps simply exposing the children to the other language would be enough for them to acquire these sound patterns and they could do it without the need for an adult who interacts with them. This was experimentally tested in a follow-up study in which another group of babies were exposed to a modified learning protocol. The difference was that the children either saw the tutor on television or they heard a recording of the tutor without visual input. The auditory information heard by the babies was exactly the same as that previous study in which the babies interacted with the adult. That is, the information that would allow them to distinguish the contrast was the same as in the previous study. The only thing that changed in the second study was that there was no social contact, no tutor present to interact with them. Would the babies now learn the phonological contrast in the foreign language?

The answer was a resounding No. That is, with only auditory exposure and without the presence of the tutor, the ability to distinguish the phonological contrast of Mandarin was exactly the same as that of the children from the previous study who had only heard English. These results suggest that social communicative interaction is fundamental to the learning of a foreign language, and that mere exposure does not seem to lead to such learning. This is because the motivation and attention of the child is much greater when interacting with someone than when being a passive receiver of information. So, if you really want your child to learn another language, interact with them in that language. Do not trust that the pictures on the TV will do the work for you. As the saying goes: *No pain, no gain.*

LANGUAGE AS A SOCIAL MARKER

Before concluding this chapter, I would like to share with you some studies that show just how important language use is from a social point of view and the consequences it may have for the use of a second language.

We cannot avoid dividing up our social contexts according to the various characteristics of others around us in relation to our own. We consider things such as skin colour, gender, the way people dress, and obviously their language. This helps us to identify with those who are like us and differentiate ourselves from those who are not.

For better or worse (worse especially if this leads to the marginalization of those who are unlike us), humans behave this way. The question that interests us here is to what extent children benefit from linguistic characteristics that influence their preferences in social interaction.

Consider the following simple yet clever experiment. A group of five-year-old speakers of English were presented with a series of faces of other children talking and asked to choose which of them they would like to have as friends. The faces appeared in side-by-side pairs. The 'catch' in the experiment was that one of the faces speaks in English and the other in a foreign language, French. Which one will they choose? It seemed that for the most part the children showed a preference for faces that spoke in English and not in the foreign language. Of course, one would assume that because the children did not understand French, they preferred to be friends with another child they understood. But there is more to it: in a second experiment the same faces were presented, but this time they spoke in English with either a native accent or foreign (French) accent. Although the children had no problem understanding the English with a French accent, they continued to choose as friends those children whose first language was English. In other words, *if you have an accent you do not belong to my group!* And when those who spoke with foreign accents were compared to those who spoke in a foreign language, the children still showed a preference for those they understood.

One of the factors that we consider when deciding on who we want to socialize with is the language those individuals speak and the accent they have in that language. But how is this determined in our social decisions? In one study, researchers explored the extent to which skin colour and language affect the decisions children make about who they want to relate to. The result was surprising. The children preferred to interact with those who have the same skin colour and accent. But what happened when these two properties were at odds? Did they choose a child with the same skin colour and foreign accent, or a child with a different skin colour and no foreign accent? It turns out that accent was the determining factor. The children preferred to interact with other children of a different skin colour as long as they were native speakers of English; more so than with children with the same skin colour but who spoke English with a foreign accent. In other words, their preferences were more determined by how the other children talked rather than by the colour of their skin.

There are other studies that support similar findings which underscore the importance of language as a social identifier. What we still do not know is whether speakers who are bilingual are more flexible in language use as a social characteristic. We will return to this question in Chapter 5.

We still have a lot of ground to cover regarding how all these factors work and how bilingualism 'tunes' the processes involved in language acquisition. Moving forward in this field will be challenging given the complexity of working with bilingual babies. As we said above, although the use of two languages in humans is increasing, it is still difficult to find societies where bilingualism is more homogeneously dispersed and where access to babies with similar bilingual characteristics is plentiful. Perhaps the greatest impediment is that this type of research is considered 'basic', which implies that the knowledge we learn from it does not have immediate application. Regrettably, this leads many people to shamelessly say, 'What's the point of all this? Why does it matter what a bilingual baby does compared to a monolingual baby?' I hope I have convinced you that it is indeed worth carrying out these bilingual studies.

Returning to *The Godfather: Part II*, Vito Corleone was not born in a bilingual cradle, but in a Sicilian one. But perhaps some of his children, Michael, Santino, Fredo, and Talia, *were* born in a bilingual cradle. The challenges they faced during language acquisition were different from those their father experienced. We know that nothing bad happened to them, at least as far as the acquisition of language is concerned.

2

Two Languages, One Brain

Even though evolution has resulted in millions of species, it still has not generated my favourite species, the Babel fish. This animal was dreamed up by the British writer Douglas Adams in his wonderful novel *The Hitchhiker's Guide to the Galaxy*. If you have not read it, put down this book and go out and buy it . . . and later, we'll see each other at the 'restaurant at the end of the universe'. Here is more about the Babel fish:

> 'The Babel fish,' said *The Hitchhiker's Guide to the Galaxy* quietly, 'is small, yellow and leech-like, and probably the oddest thing in the Universe. It feeds on brainwave energy received not from its own carrier but from those around it. It absorbs all unconscious mental frequencies from this brainwave energy to nourish itself with. It then excretes into the mind of its carrier a telepathic matrix formed by combining the conscious thought frequencies with nerve signals picked up from the speech centres of the brain which has supplied them. The practical upshot of all this is that if you stick a Babel fish in your ear you can instantly understand anything in any form of language. The speech patterns you actually hear decode the brainwave matrix which has been fed into your mind by your Babel fish.'*

Don't tell me that the Babel fish isn't an interesting species! How many problems could we solve if this creature really existed . . . the strangest creature in the universe. At the very least, we wouldn't have

---

* Douglas Adams (1979), *The Hitchhiker's Guide to the Galaxy*. London: Pan Books, pp. 40–41.

to worry about struggling through second-language learning. We would just go to a fish store – problem solved.

Bilingual speakers are not Babel fish (they are not used to whispering in anybody's ear), but they do have something in common: in both of their brains there must be linguistic representations that correspond to two languages. In other words, the only way the fish can translate from one language to another is by having both of them stored in its little brain. And while bilinguals only speak two languages and not *all* languages in the universe, the question is the same: how do two languages coexist in one brain, and what are the consequences for their continuous use? This chapter is dedicated to this issue and others intrinsically related to it.

The study of how the brain sustains higher-level cognitive abilities, or what we will refer to as cortical representations of cognitive functions (language is one of them), is extremely complex. The brain and cognitive bases of language, memory, attention, emotion, and so on, are difficult to study. This is because, among other things, the cognitive processes involved in these capacities are not independent, but interactive, and in complex ways. Think about how the emotional system interacts constantly with the attentional system when a highly emotional stimulus suddenly piques our interest. Remember, for example, the last time you were at a noisy party and were trying to have a conversation. You could probably barely pay attention to the person speaking to you and all the other conversations around you just seemed like background noise. However, if someone having a nearby conversation had said your name, perhaps that would have caught your attention. So, even though everything but your own conversation seems just like noise, your ears would have detected your name and would have directed your attention to that conversation. Yes, our name is a highly emotional stimulus: we care very much about what other people say about us.

To make things more difficult, the more we understand about the relationship between cognition and the brain, the more it is evident that higher-level cognitive functions involve neural circuits that are distributed in different areas of the brain. This is not to say, however, that there may not be certain areas that have a greater or lesser importance in the functioning of each of these skills, but it does mean that the relationship between the brain and cognition is even more complex than we thought. You can think of the brain as an orchestra and as in any orchestra, there are different instruments with greater or lesser importance to mark harmony, melody, or rhythm in a musical piece.

For many years our knowledge about how language is represented in the brain has come from studying the verbal behaviour of people who suffer some kind of brain damage. We call these language disorders *aphasia*. Brain damage can arise from various causes, such as tumours, infections, congenital malformations, strokes, neurodegenerative diseases, or traumatic brain injuries. The study of how injuries in different areas of the brain result in different verbal behaviour patterns has been fundamental to relating cognitive functional models of language, informed by linguistics and cognitive psychology, with neural correlates. However, in the last thirty years, the development of neuroimaging techniques has dramatically advanced the field of cognitive neuroscience. These techniques allow us to 'see' live (or almost live) brain activity of healthy people while they perform different experimental tasks. For example, we can analyse which brain circuits are activated when reading a text compared to naming drawings, hearing phrases, or thinking about plans for the weekend.

We can register the brain activity triggered by these tasks by measuring the oxygen consumption of certain areas or by registering the electrical activity generated by groups of neurons. In addition, the degree of temporal and spatial precision is more than adequate. These techniques also allow us to make predictions about which areas of the brain should be most involved in different aspects of language processing. These hypotheses were more difficult to make when we were able to study only the verbal behaviour of people with brain damage, and, in many cases, we could only know with certainty which tissue was damaged after the patient's death. Let's see how these studies have helped us to better understand how two languages coexist in the same brain.

BRAIN DAMAGE AND BILINGUALISM

In one of the 2015 Formula 1 World Championship pre-season training sessions, the McLaren driver Fernando Alonso had an accident: he hit the wall of a curve in the Montmeló, Catalonia, circuit. As a result Alonso suffered a concussion and had to be admitted to the hospital, where he was kept under observation for a couple of weeks. Fortunately, he recovered successfully and continued competing in the world championship. The causes of the accident still have not been completely clarified. At first glance, it seemed strange that a driver as experienced as Alonso had made a mistake, apparently an enormous one, which led to all kinds of speculation regarding a technical failure of the car. I'm not a frequent follower of motor sports, so I wouldn't have paid attention to this if it had not been for the following: along with the rumours about the cause of the accident, news began to spread that right after the accident Alonso could only speak Italian (a language he knew and used often, among other things for having been a former driver of the Ferrari team), but not his native language, Spanish, or the language with which he interacted daily with the members of his team, English. This was framed as strange behaviour on his part. There were headlines such as 'Fernando Alonso Wakes Up in Italian' and some that even surprised me like 'Alonso Not the First Spanish Athlete to Wake Up Speaking in Italian' (in case you are interested, the other was the cyclist Pedro Horrillo).

I find this story especially interesting for two reasons. First, because Alonso's strange verbal behaviour made so many people (or at least journalists) pay attention, showing that there is general interest about language, and in this case, about bilingualism. In fact, this news attracts attention regardless of whether the affected party is well known to the public, as we can see from the story of an American man who woke up from unconsciousness speaking Swedish.* It's funny how these cases often lead to bizarre speculation such as wondering whether the man knew Swedish before losing consciousness or

---

* 'An American Wakes Up with Amnesia Speaking Swedish', *La Vanguardia*, 17 July 2013.

whether his ancestors were Swedish. In any case, we probably agree on the following: no brain damage can result in the sudden learning of a new language, nor can the knowledge of a language be transmitted through genes, at least as far as we know right now.

The other reason why I find Alonso's case especially interesting is because he himself denied that the situation had taken place. In subsequent statements the driver said: 'It was all normal, I did not wake up in 1995, or speak in Italian, or anything that has been said. I remember the accident and everything that happened.' Who knows why anyone would say that Alonso woke up speaking only in Italian for a few minutes.

Before we explore what we know about the linguistic deterioration of a bilingual's languages resulting from brain damage, we should take a little time to define some basic concepts in neuropsychology.

The first lesson I learned from Alfonso Caramazza during my post-doctoral work at Harvard University was that in neuropsychology there are two types of behavioural patterns that are highly informative. On the one hand, we have the so-called *associated deficits*, which refer to two or more linguistic dysfunctions that appear together as a result of damage to a specific area of the brain. For example, if a bilingual speaker shows a specific dysfunction in each of his languages due to this damage (for example, he has problems repeating words), we refer to this as an association of dysfunctions in the two languages. That is, the two languages are affected in the same way by the brain damage. On the other hand, more interesting, perhaps, is what *dissociated deficits* tell us. Imagine in this case that, as a result of brain damage, a person shows certain linguistic problems in one of his languages but not in the other. In other words, the patient shows a dissociation between speech in his two languages. That is, his ability, for example, to repeat words in one language is dissociated from his ability to repeat words in the other language. Dissociations offer us a lot of information, because they suggest that whatever brain damage a patient suffers, this affects one type of cognitive process (repetition of words in language A) and not other types (repetition of words in language B), which in turn suggests that such processes are supported by different brain circuits that are to a certain extent cognitively independent.

Maybe the following analogy will help. The windshield wipers of a car are independent of the braking system. So we may encounter situations in which one works but the other doesn't. However, both rely on the correct functioning of the electrical system and, therefore, if that breaks down, both the windshield wipers and brakes will stop working. In the first case, there is a dissociation and in the second, an association. Below I describe an example of these dissociations, which we will return to later in cases of bilingualism. (If you are curious to learn more about these dissociations, Oliver Sacks has written very well about them.)

Students in secondary education tend to find language classes difficult and boring, especially when it comes to syntactic analysis. In this case, they may be right: the subject is indeed difficult and, sometimes, boring. If you had to study language this way, you might remember generating tree structures based on sentences (subject, predicate, etc.). To create one of those trees, you had to determine the grammatical categories of the words and their function with respect to the rest of the words within the sentence. It is much more difficult to do this in terms of language use, at least with regard to spoken language. However, not everything is that complicated, and one of the things that children learn naturally is the difference between nouns and verbs. Determining what nouns and verbs are is extremely simple compared to identifying determiners, adverbs, and conjunctions, for example. It is as if we understood the relationship between objects-nouns and actions-verbs naturally. And, in fact, the linguistic difference between nouns and verbs exists in all languages and is a central grammatical property in linguistic theories. This difference reflects, to some extent, our view of the world or conceptual structures: nouns tend to describe objects and verbs tend to describe actions. They tend to, just tend to. In addition to the fact that this difference is useful in linguistic descriptions, the question that we are interested in here is to what extent this difference has a cerebral correlate. That is, it is not immediately obvious that there are certain neural circuits that facilitate to a greater extent the processing of nouns and others that support the processing of verbs.

As it turns out, there are quite a few people who after brain damage have more problems processing nouns than verbs. In addition, the opposite pattern can be found in other patients, that is, those who experience more problems with verbs than nouns. Some suffer from what is called *anomia*, which refers to the difficulty in accessing words from the mental lexicon when we want to express ourselves. To put it simply, these people suffer much more often than healthy speakers from situations of having something 'on the tip of the tongue'. Imagine how cumbersome that would be! When these patients are asked to say aloud the name of an object in a drawing (like *a broom*), it is common for them to fall into an anomic state in which they cannot recover the name of the object, although they know perfectly well what object is represented in the drawing. But interestingly, it is in that same situation that the patient may be able to name the verb that corresponds to the action that is carried out with that noun (like *to sweep*). In other words, brain damage can affect certain grammatical categories' representations to a greater extent than other categories' representations – what we have previously described as a dissociation of deficits. These observations suggest that, in fact, the difference between nouns and verbs does not only have implications for linguistic theories, but our brain seems to take this difference into account when organizing the mental lexicon. We will return to this example of dissociation later.

The question that interests us now is to what extent a brain lesion affects each of the bilingual speaker's languages differently, and whether we can observe some sort of relatively constant pattern. My opinion on this matter is perhaps somewhat controversial, but I believe that the more common pattern, and by far, is that the two languages of the bilingual speaker are each affected in a very similar manner and degree. In other words, it does not seem very common to find cases in which one of the languages is much more affected than the other, always considering, of course, the degree of knowledge of the two languages prior to brain damage.

I say that this view is controversial because you will be able to find a long list of different patterns of impairments and recovery of the two languages in other books on bilingualism and neuropsychology. For example, in the typology described by Michel Paradis, we find up to five types of linguistic recovery: the pattern of *parallel recovery* is when the patient recovers his linguistic abilities similarly in both languages; *differential recovery* occurs when a patient recovers one of his two languages to a level similar to that which he had before the brain damage but does not do so for the other language; the *antagonist recovery* refers to a curious situation in which the recovery of one language negatively affects the other language. Finally, there are two more typologies. *Successive recovery*, where one of the languages begins to recover only when the other is fully recovered; and the so-called *blended recovery* in which both languages are mixed involuntarily, thus hampering their restoration.

I'm not saying that these types of cases do not or cannot exist, or that they are uninteresting (in fact, I think they are quite interesting, as we will see later). All I'm saying is that the most common case is parallel impairment of both languages. In addition, the examples that support some of these dissociations are more often from clinical observations (often when the patient's impairment is acute) than from controlled, systematic studies. However, I must admit that there are contradictory findings. I think it is relevant to mention here that studies on bilingual aphasics are especially complicated, given that, very often, it is difficult for us to know precisely what the patient's linguistic levels were before impairment and what he did with his languages. To further complicate the matter, factors such as age of acquisition of the second language and linguistic dominance (the language that a person uses with more fluidity) can also affect impairment and recovery patterns.

From my point of view, there are two reasons why the most frequent pattern of language deterioration is parallel. The first is that, as we will see later, neuroimaging studies show that there is significant overlap between the areas of the brain that sustain the processing of both languages. So, if there is such overlap, at least at the macroscopic level, it is reasonable to think that the two languages are affected similarly in many cases. The second reason has to do with the fact that oftentimes linguistic impairments are the result of damage affecting several areas of the brain, making it difficult to detect potential dissociations between languages. Therefore, in principle, it is possible that certain neural circuits are more involved in the representation of one or another language, but that these differences are only visible in a microscopic way.

Here are a couple of examples about the type of evidence that supports parallel impairment of the two languages. The first comes from a study that came up as a result of a question that my mother asked me one Sunday afternoon while we were eating. The question was simple: 'I have a friend who has been diagnosed with Alzheimer's disease. Although I've always spoken to her in Catalan, which is her second language, in what language will I likely end up speaking with her, in Spanish or Catalan?' What my mother asked was a common concern. Let's reformulate the question in academic terms: how do languages deteriorate as a consequence of a neurodegenerative disease? My answer was even simpler: I did not know, and the worst thing is that there were not many studies I could consult.

After consulting the few studies carried out on the subject, I noticed that the issue was not entirely clear, so I set out to do something about it. In collaboration with neurology departments from several hospitals in Barcelona, we evaluated the linguistic competence of three groups of Spanish-Catalan bilinguals. These bilinguals had spoken on average for more than fifty years in the two languages and possessed a high knowledge of both. Most of them lived in the metropolitan area of Barcelona, where the daily use of both languages is very common. Two of the groups consisted of participants who had been diagnosed with Alzheimer's disease and were in a mild to moderate stage of the disease, according to standardized neuropsychological tests. The third group included individuals who suffered from mild cognitive impairment and those who had not been diagnosed with Alzheimer's. In different experimental sessions, we asked the participants to say the name of what was presented in drawings in both their languages. They also performed a translation task in which they were presented with a word in one language that they had to say aloud in the other language. As you can see in Figure 2.1, we obtained at least two clear results. First, bilinguals who had poorer performance in neuropsychological tests also had poorer performance in the linguistic tasks that we had designed. This was not a very surprising result since it is expected that an impact on the cognitive system in general also harms language. Second, the language impairment associated with cognitive deterioration was of equal magnitude for both languages. Although the participants performed the tasks a little better in the language they reported as more dominant (whether it was the first they had acquired or not, and whether it was Spanish or Catalan), there was a pattern of parallel language impairment. In addition, the type of errors that participants committed in both languages was also similar. For example, although the percentage of errors arising from an unintended language (a translation into the irrelevant language) was greater for the non-dominant language, the pattern of deterioration was also parallel. In other words, the disease was deteriorating the two languages in the same way and at the same pace. I could now tell my mother that her friend would continue to speak the same language but with greater difficulty.

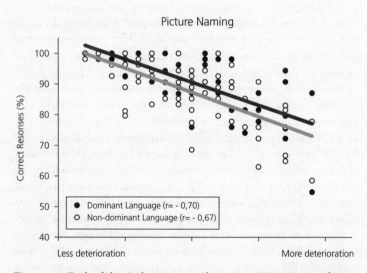

Figure 2.1. Each of the circles corresponds to a participant's score from the three groups in the study. Black circles correspond to the score in the dominant language and white circles show scores in the non-dominant one. The horizontal axis of the graph corresponds to the participants' scores in a standard neuropsychological test. The further to the right the circles are, the greater the participant's cognitive impairment measured in that test. The more cognitive impairment there is, the worse the scores are. The slope is similar for both languages, reflecting parallel impairment.

The second example has to do with the impairments that we commented on earlier, where there are much greater difficulties in accessing words from one grammatical category compared to another. Remember: verbs before nouns. We only have a partial understanding of the origin of this dissociation and how the brain represents words or lexical items. However, the phenomenon itself would seem to indicate that, at a minimum, the grammatical distinction of noun versus verb is one of those that the brain considers when organizing lexical information. In the context of bilingualism, the question is whether the two languages are organized based on the same variables or dimensions. About eight years ago, we had the opportunity to address this question, when a fifty-five-year-old bilingual man who suffered from a progressive primary aphasia was generous enough to collaborate with us by participating in several linguistic tests. His disease is a neurodegenerative one in which one of the most notable symptoms is the progressive deterioration of linguistic capabilities beginning from the earliest stages of the disease.

Thanks to this man's patience, we were able to track his linguistic skills for two years, which allowed us to evaluate how language skills were deteriorating as the disease progressed. From our own observations, we quickly noticed that he had many more problems when the activity involved verbs than when it involved nouns. His mistakes were basically due to anomic episodes (tip-of-the-tongue states), but he also made some semantic errors, for instance saying 'pear' when shown a drawing of an apple. In addition, he showed worse performance overall in his second language (Catalan) than in his dominant one (Spanish), despite having learned them both before the age of four, and having used Catalan with his wife and children. What was more interesting was that the dissociation between nouns and verbs was present in both languages. This was not an isolated case and, in fact, it perfectly complemented our observations from a previous case in which a patient who suffered from Alzheimer's disease showed the opposite dissociation, that is, a much larger and disproportionate impairment for words corresponding to nouns compared to verbs, but again, in both languages.

These cases, and similar ones, suggest that the brain tends to apply the same principles to both languages when it organizes information about words – in this case, their grammatical category. In other words, the properties that are important for language organization in the brain are the same for a bilingual's two languages. And in fact, this claim is consistent with the results of studies in which brain activity is analysed during the processing of nouns and verbs among healthy bilinguals. In these studies, it has been found that certain cerebral areas seem to have different degrees of influence on the representation of these different grammatical categories. The crucial point to mention here is that these differences are also observed in the second language.

These studies are just a few examples of the many investigations which show that parallel impairment in the two languages is the most common pattern of linguistic impairment due to brain damage. However, there are also other studies that show some dissociations.

Consider, for example, this clinical case: Raphiq Ibrahim from the University of Haifa studied the verbal behaviour of a forty-one-year-old man who suffered a brain injury as a result of encephalitis caused by a type of herpes simplex (yes, the same one that can appear on the lips can spread to the brain and severely affect it). This injury especially affected his left temporal lobe, an area of the brain critical for language processing, among other things. The patient was a high-school biology teacher in the city of Haifa, Israel, and although his first language was Arabic, he was highly proficient in Hebrew, a language that he had learned at the age of ten and used regularly both at school and in his private life. Ibrahim explored the patient's verbal behaviour on several linguistic tasks in his two languages two years after the patient suffered the injury and after he underwent surgery to remove the damaged area. The patient had little speech fluidity, with many pauses and anomic states in which it was difficult for him to access words from the mental lexicon. However, reduced fluency was much more evident when the patient spoke in Hebrew than when he spoke in Arabic. Although his scores on standard tests, which involved naming drawings, were below normal in both languages, they were much worse in Hebrew. Also, his understanding of speech and the ability to read and write were more impaired in Hebrew. Nevertheless, and interestingly, his ability to simply repeat words was not affected in either language. The patient received intensive language therapy in both languages for three months, and even though his performance got better in both, such improvement was much more apparent in Arabic. These results led the author to argue in favour of the existence of cortical centres that are specific to each of the languages, in this case, two similar languages that are both Semitic.

Before moving on to the next section, I would like to take a moment to thank all the people and their families who have collaborated in this type of research, always with a commendable predisposition to help science. This help is especially generous when one is suffering from cognitive deterioration due to a disease. Indeed, they are patients with great strength and courage. To all of you, thank you, truly.

PHOTOGRAPHING THE TWO LANGUAGES

Almost twenty years ago, while I was doing my PhD, I was a research assistant in one of the first studies to be carried out with the objective of exploring how two languages were represented in the bilingual brain. The study aimed to investigate the effect of the age of acquisition of the second language on both languages' cortical representation. This involved analysing brain responses of highly proficient bilingual speakers using positron emission tomography. To do so, we studied Italian-English bilinguals with a late age of second-language acquisition (at age ten) and Castilian-Catalan bilinguals with an early age of second-language acquisition (at age four). One of the difficulties we faced was that at that time our laboratory did not have access to that particular neuroimaging technique, so, in collaboration with a team of neurologists from Milan, we decided to conduct the experiment in the Milanese hospital of San Raffaele. This meant that the participants had to travel from Barcelona to Milan, in such a way that they also had a pleasant end of week in the Lombard city. It was all in the name of science.

There have been many studies exploring the brain activity of bilinguals while processing their two languages. Work has been carried out using various techniques (functional magnetic resonance imaging, positron emission tomography, magnetoencephalography, and so on), experimental paradigms, and language pairs. It would be hard to describe them all here, but instead I will try to give a general overview of what I think we have learned from them.

At a general level, we can say that the areas of the brain involved in the representation and processing of a bilingual's two languages are the same. It's as if the brain were somehow prepared to handle any language signal in the same manner regardless of the language or languages to which it is exposed. However, this does not mean that there aren't certain differences in their cortical representation, which will depend on many variables, such as the age of acquisition of the second language, the proficiency level of this language, and the similarity between the two. To make things more difficult, these variables tend to interact in complex ways. This is just a rough generalization but let's dig a little deeper.

Let's look at, for example, the following meta-analysis which compared the results of fourteen studies that used functional magnetic resonance imaging to explore the cerebral representation of bilingual speakers' two languages. The authors separated the studies according to the degree of knowledge (the competence) that the participants had in their second language. In eight of these studies it was considered that the participants had a high command of the second language, and in the remaining six the participants were deemed to have a moderate to low command. In the first of these subgroups, activation was detected in the left hemisphere in the classic brain network that is involved in language processing, including fronto-temporal regions.

In Plate 1, the areas in red correspond to the activation of the dominant language, the areas in blue represent activation of the second language, and purple shows areas that are activated when both are processed. In panel A, which depicts highly proficient bilingual speakers, there is a large overlap in this network between the two languages. Indeed, nearly all the colours in panel A are purple, that is, almost all those areas that correspond to the first language also correspond to the second, and vice versa. On the other hand, the results of the studies that involved moderate- to low-proficient bilinguals were somewhat different. As you can see in panel B, there is much less overlap between the two languages. Let's pause for a moment to analyse these differences.

At first glance, it seems that the second language is represented in a more distributed network than the first language, that is, it tends to involve more areas of the brain. Also, when comparing the activation triggered by the second language in bilinguals of different competences, the less-proficient bilinguals seem to require more right-hemisphere areas, as if this were a compensation mechanism. This is interesting, because there is clinical evidence showing that lesions in areas of the left hemisphere (especially frontal areas) can force corresponding areas of the right hemisphere to perform, to a certain extent, the left hemisphere's functions.

Another interesting finding is that the left superior temporal gyrus was less activated for the bilinguals with lower second-language proficiency. This area of the brain has been linked to conceptual or semantic processing. One possible interpretation is that this is less activated when knowledge of a language is less. That is, the semantic information that we extract from a second language in which we are not very proficient is less than what we extract from our native language. It makes sense. Less competence in a second language also results in greater activation of areas related to language control, such as the dorsolateral prefrontal cortex and the anterior cingulate cortex, which could be interpreted as a greater need for attentional resources when encountering the second language. We will return to this question in various sections. In short, these results suggest that processing a second language in which one is not very competent is costlier and, consequently, the processing of a second language requires a more extensive brain network.

It is possible that for the last few paragraphs you may have been thinking that the level of competence in a second language is usually linked to when it was learned. Although this is not always the case, when a second language is acquired as a child and, more importantly, continues to be used, the odds are that the speaker is highly competent in it. The question, then, is to what extent the cortical representation of this second language depends on its age of acquisition and not just on its level of competency. As it turns out, just like the competency level in a second language, the age of acquisition also seems to have independent effects on cortical representation. For example, in tasks that involve semantic and grammatical processing, such as understanding sentences, a language learned relatively late (during puberty or later) tends to activate areas related to language, such as Broca's area and the insula, to a greater extent than the first language. In fact, this latter finding is consistent with others that reveal that, in the first language, words that are acquired later (for example, 'screwdriver') produce greater neuronal activity than words learned at younger ages (such as 'rabbit'), particularly in areas related to phonological processing and motor planning of speech. These differences between languages do not seem to be present when the two languages have been learned in the first years of life and competency in both languages is high.

Given this scenario, it is reasonable to ask the following question with respect to the cortical representation of a second language: what has a greater effect, the age of acquisition or the acquired competency? That is, what has more influence on how the brain represents a second language, having learned it as a child or knowing how to use it extremely well? It is difficult to answer this question because, as we have seen, there is an important correlation between the two variables and, therefore, it is difficult to evaluate their effects independently. Moreover, these two factors can influence different linguistic aspects in various ways. For example, it has been suggested that semantic or conceptual processing in the two languages is very similar among individuals who have attained a high level of competency in the second language regardless of their age of acquisition. However, when syntactic processing is measured, there seem to be certain differences, and the age of second-language acquisition has important effects that are independent of the level of competency. It is still difficult to say which of these variables has a greater effect on the cortical representation of a second language.

There are several explanations as to why brain activation is greater when processing a second language compared to the first, especially when the level of second-language proficiency is not very high. There are various factors that are not mutually exclusive, such as the cost associated with controlling two languages, the lack of automaticity in processing a second language, the cognitive effort that this may entail, and the greater burden on second-language motor control.

INTERFERENCE

As we have seen, neuroimaging techniques are allowing us to discover the cerebral bases of linguistic processing both in bilinguals and monolinguals. Using these techniques, we can identify areas of the brain that are involved in certain linguistic activities. However, they also have some limitations: among other things, they do not allow us to identify areas of the brain that are 'essential' to carrying out a specific task. Let me explain: it's one thing that a part of the brain is activated while performing a specific task (for instance, processing a second language); it's another thing that this activation is essential to perform this task. Let's go back to the analogy between the brain and an orchestra: imagine that we are listening to a concert in which a violin has a solo but there are other instruments that enter and accompany the violin such as a tuba, drums, and so on. The piece will sound very good if all of them are playing their part. In fact, to a non-expert in music, it may seem that all the instruments are equally necessary for the piece of music to make sense and sound good. However, while the violin solo plays a fundamental role, the tuba may not. So if the tuba part weren't played, the concert would continue sounding 'relatively good'. But without the violin, the result would be much worse.

To discover what areas of the brain are essential to carry out a task, we have to look at what happens when those areas do not act correctly, either because they are damaged (as we've seen before) or because we interfere with its normal functioning. Currently, two of the most commonly used techniques to interfere with the operation of certain areas of the brain are transcranial magnetic stimulation and intraoperative cortical electrical stimulation.

Transcranial magnetic stimulation uses a metal coil to generate a magnetic field that is applied on the skull of an individual. The magnetic field in turn produces an electric field in the brain that interacts briefly with the normal electrical functioning of neurons. Don't be alarmed: the stimulation is painless and the neuronal alterations are temporary, such that the neurons return to their normal state after a very short period of time. What this technique allows us to do, then, is to alter the normal functioning of cortical structures. Stated in a more exaggerated way, it gives us the ability to produce virtual, brief injuries (and sometimes enhance brain functions) in healthy individuals, and to analyse what the result is from a behavioural perspective. This is important because it allows us to establish causal relationships between the neurons stimulated and the cognitive functions they produce. Importantly, this technique is also used for therapeutic purposes in cases of depression, migraines, epilepsy, and so on. Currently, the number of studies that have explored language representation in bilingual individuals using this technique is limited. However, results show that the temporary interruption of certain areas of the brain (such as the prefrontal cortex) may result in a lack of linguistic control, which may cause a person to involuntarily mix languages or even block access to one of them to a greater or lesser extent. For example, the stimulation of the dorsolateral prefrontal cortex causes problems when choosing a language and avoiding interference from the other language. It is as if the speakers subjected to stimulation in that area had lost control of the languages they know. In the next few years, we will surely see a boom in studies of this type in the context of bilingualism.

Let's move on to the second technique mentioned above, cortical electrical stimulation. I couldn't resist sharing with you a figure of the so-called cortical homunculus, or Penfield homunculus, that appears frequently in neuroscience textbooks. As you can see in Figure 2.2, the homunculus is a representation of our entire body in the brain, both in terms of sensitivity and motor skills. And yes, this map really exists, even if it looks a little like a cartoon.

How was this map created? Well, by using intraoperative cortical stimulation. When employing this technique, if certain areas of the brain are activated by electricity, you can determine their relationship to certain capabilities. It's possible to come up with a somatotopic, motor, or cognitive abilities (e.g. language) map, thanks to pioneering studies carried out in the 1950s by, among others, the neurosurgeon Wilder Penfield. Today, this technique is always used for medical purposes: for example, when neurosurgeons have to remove a brain tumour and need to know what the side effects of such surgery will be on the patient. Depending on the location of the tumour, one of the cognitive abilities that is 'mapped' is language, since it is a

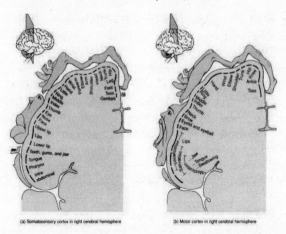

Figure 2.2. The cortical homunculus, showing the anatomical divisions of the primary somatosensory cortex in image (a) and the primary motor cortex in image (b) (taken from http://personal.uwaterloo.ca/ranney/mindheartsoul.html).

fundamental function that the neurosurgeon should avoid damaging in the course of the operation. But how can the neurosurgeon know what stimulation does to language processing? The procedure of cortical stimulation is performed while the patient is awake. Once the surgeon has opened the skull and has accessed the brain, general anaesthesia is reduced and the patient is revived while the surgeon continues to apply local anaesthesia on the scalp and skull. The surgeon can apply electrical stimulation directly to the brain without producing pain, since the brain does not have pain receptors. The patient is then asked to, for example, name what he/she sees in a series of drawings while electrical stimulation is applied to the different areas that could be damaged in the operation. The stimulation will affect the patient's ability to carry out this task only in some areas (it would be like taking away instruments one by one from the orchestra and seeing how the musical piece sounds). In this way, if they were compromised by surgery, the patient could end up with problems with language use, which would dramatically affect their communicative capacity. It's better to leave these areas alone.

Can we have a mapping similar to that of the homunculus for both languages of the bilingual? That would be nice, but this is more complex. It is interesting, however, to see that this question was already intriguing to Dr Penfield, who resided in a bilingual area, the region of Quebec. In an interview with the Canadian newspaper *The Montreal Gazette*, Dr Penfield answered a question regarding the convenience of education in two languages. The interview was titled 'Bilingual Brain Superior – Penfield'. This fact wouldn't be very surprising if it weren't for the fact that it was published on 15 June 1968 . . . more than fifty years ago. And the debate continues! Why might this be the case? When the patient is a bilingual speaker, a mapping is often done of the two languages, so as to know which areas, upon stimulation, interrupt the processing of both or one of the languages. The results of these investigations are somewhat contradictory. While there are studies that show a wide overlap between the areas of the brain responsible for processing the two languages, others have identified that the stimulus only affects one of them. When this is the case, in general, it seems that there are more areas involved in processing a second language compared to the first. It appears that the dominant language requires fewer neuronal resources for processing.

Let's consider a study conducted by Timothy Lucas and his collaborators at the University of Washington and published in the *Journal of Neurosurgery*, in which the areas of the brain that interfered with picture naming were mapped in the first and second language among twenty-two patients with epilepsy. In twenty-one of the patients some areas of the brain were identified that interfered specifically with either the first or second language. However, it is important to note that fewer than half of the patients showed common areas of representation for the two languages, areas that, when stimulated, interfered with the processing of both languages. Finally, this study also compared the linguistic organization of bilinguals with that of 110 monolinguals and, as expected, similar results were found. Together, these results were interpreted by the authors as follows: there seems to be some functional separation between the cortical representation of the two languages. That is, there are areas of the brain that are fundamental for the processing of the first language and others for the second. Likewise, there are certain areas that are involved in the processing of both languages. Finally, the representation of the first language in bilinguals seems to be similar to the first language in monolinguals, which would suggest that learning a second language does not significantly alter the cortical representation of the first.

Studies of this type allow us to have more accessible and precise information about the areas of the brain involved in, or rather essential to, cognitive processes. Although they could be considered opportunistic, because they must always have medical implications, I think they will uncover very relevant information in the coming years. In particular, the recording of electrical activity and brain stimulation with implanted electrodes offer the possibility to explore the verbal behaviour of patients more exhaustively. These implants also meet medicinal needs, and are usually implanted to explore the origin of epileptic seizures of patients that have not responded well to conventional pharmacological therapies. Be on the lookout for more about this matter in the newspapers.

CONTROL, CONTROL, CONTROL

If you've ever tried to learn a foreign language, you have probably experienced the unpleasant sensation that, when you finally have enough courage to try to address someone in that new language, the words don't come to mind. There is no doubt that you know what you want to say and even know the words that you want to say; but, when putting them together and opening your mouth, things just don't seem to flow. You may also have the sensation that if you decide to try to speak anyway, the words will come out one by one without forming coherent sentences, or you will notice a massive interference from your dominant language. Do not get too frustrated; it happens to everyone. You probably already know that, bad as they may seem, situations like these are what lead many people to believe that they are better at understanding than speaking a second language, a statement that in many cases I think illustrates our perception of linguistic understanding rather than reality.

These fluency problems arise, in part, because it is difficult for us to control access to the second language. Not only is it hard for us to access its words and grammatical structures, but the words and grammatical structures of the first language are also there, interfering with our verbalization. Recently, a friend gave me an excellent example of this interference. In Barcelona, all tourists want to visit the Basilica and Expiatory Church of the Holy Family (*la Sagrada Familia*), and my friend kindly helped a clueless group find it by giving them directions in English. The tourists were grateful and gave her a sober 'thank you', to which my friend responded with a very polite '*de* nothing'. '*De* nothing?!' This wasn't due to lack of knowledge, it was lack of control. Of course, my friend knew that this was not a phrase in English, but she also knew what words to say to return the courtesy; in fact, another simple 'thank you' would have sufficed. But this time, her tongue did not obey her brain.

My friend is not alone in situations like this. All of us who have tried to master a second language in adulthood have realized that it does not only involve learning new linguistic representations, but it also requires the acquisition of a special skill we call 'linguistic control'. This is fundamental to being able to acquire the verbal fluency that allows us to communicate efficiently, and to say 'you're welcome' instead of '*de* nothing'. But how is linguistic control acquired? Well, you know, it's like grandma's special ingredient: practice.

Bilingual speakers who are competent in two languages are like jugglers. When the communicative situation requires it, they are capable of focusing their speech on one of them without apparent difficulties, avoiding the massive interference from the other language's representations. So, for example, if an English-Spanish bilingual is interacting with an English monolingual, he will rarely have intrusions from the Spanish lexicon or commit a translinguistic error, that is, the error of 'slipping' a word from Spanish into the conversation in English. Think about it: if this were common, communication with bilingual speakers would be impossible (unless we knew their two languages), and bilingualism would clearly entail problems for communication. That is to say, if at all times we were involuntarily mixing the lexical, syntactic, and phonological representations of two languages, it would be very difficult to hold a conversation.

Whenever I point this out, someone usually notes that there are many situations in which bilingual speakers change languages during a conversation, introducing elements from both. This is true, and we call this phenomenon 'code-switching'. This verbal behaviour, however, is far from random and does not seem to correspond to a failure of linguistic control (at least in most cases), but to other questions that are communicative in nature. What is specifically interesting to me is that code-switching adheres to certain grammatical restrictions and, therefore, cannot be considered the result of errors in linguistic control, at least on most occasions. In other words, the switches follow systematic rules. Consider, for example, the following: '*No sé dónde he dejado las keys*', where the article '*las*' matches the word 'keys' in number (e.g. plural).

Bilingual speakers are not only able to focus their attention in the desired language, but they are also capable of maintaining bilingual conversations with diligence. This concept is a bit difficult to understand if one has never experienced it before. In fact, it surprises and irritates many people who live in monolingual environments. Imagine the following situation: a family of five is eating at the dinner table (on the menu are tuna croquettes and green beans). The father speaks in Spanish with his wife and his son, but he uses Catalan with his daughter. The daughter, in turn, speaks Catalan with her father, but Spanish with the rest. The son and the mother understand the two languages, but speak in Spanish with the rest of the family, including the grandmother, who only speaks Spanish although she understands Catalan. This communicative situation is what I call bilingual conversation, in which the two languages are continuously put into play in an orderly manner. That is, it is not that they are randomly used or mixed without rhyme or reason. On the contrary: the language that is used is determined by the person to whom it is directed. We will not discuss here how these differences arise with respect to the language each individual chooses, since the reasons may be multiple and due to several causes (for example, the presence of other relatives who do not understand one of the languages). As strange as it may seem, this situation of 'orderly mixing' commonly occurs. (And yes, the family I just described is the one in which I grew up.)

At first glance, bilingual conversations present a paradox. Since all the participants at the dinner table know the two languages, wouldn't it be easier and less taxing to decide what language to speak in rather than switching between them? And if there is disagreement at the time of choosing, they could use the languages on alternate days, simple as that; this way, everyone uses both languages without problems. Well, it is not that easy, and it turns out that having conversations of this type does not seem to be all that difficult, at least for highly proficient bilinguals. In fact, it would seem that when we establish what language to use with each individual, what indeed seems difficult is to address him in the *other* language. If you do not believe it, and you know two languages, try to have a conversation with a friend in the language you do not usually use with him and see how long that lasts. So, it seems that it is more difficult to change the language in which we are used to talking to someone than to switch from one language to another depending on the listener, even within the same conversation. When we are used to talking to someone in a specific language and are forced to use a different language, we sometimes unintentionally slip into the language that is usually used with that person. For example, in a situation in which two friends use English and Spanish at the same time, when they also include someone who only knows English, they will switch to English (not only to be polite, but to be able to communicate). However, there are times when the two friends interact in Spanish, which can lead to an uncomfortable situation. Although it's hard to believe, in most cases this change is involuntary and is not meant to exclude anyone from the conversation. In fact, this may even occur among monolinguals. For example, Spanish-speaking monolinguals may say '*Encontrémonos en el* check-in' ('Let's meet at the check-in counter') or '*¿Has traído tu smoking para la cena formal?*' ('Have you brought your tuxedo jacket for the formal dinner?'). Should they not say *check-in* and *smoking* in English? This would be very difficult to change given that *check-in* and *smoking* are most frequently used.

Everything I have just explained shows that bilingual speakers can be viewed as jugglers, since they use their two languages in a quite sophisticated way. When the conversation requires it, they are able to focus on one language and avoid mixing the two while at the same time they can change from one language to another when the conversation involves bilingual situations. How do they control the two languages?

Although studying the cognitive processes and corresponding neural bases involved in linguistic control has always drawn the attention of language students, this interest has grown at spectacular rates in the last twenty years. The first issue that needs to be determined is what happens to the representations of the language that is not currently involved in a conversation, the one we call *language not in use*. For instance, when a Spanish-English bilingual is speaking with someone in English (the language in use), what happens to the Spanish (the language not in use) representations? If linguistic control acted as a simple switch and the intention to speak in a specific language was enough to 'turn off' the unwanted and 'turn on' the desired one, then this question would be relatively trivial. Simply, the system would block the activation of the language not in use and the bilingual would become a 'functional monolingual'. The reality seems to be somewhat more complex and numerous studies have shown parallel activation of the two languages, regardless of the one that is being used. Let's look at one of these studies that I think has a great deal of value.

In the study, Guillaume Thierry and his colleagues at the University of Bangor in Wales analysed whether there was activation of representations of the language not in use when bilingual participants carried out a task in the other language. In other words, whether the language that was not being used was 'turned off' or 'kept on'. The task was simple: two words were displayed on a computer screen and participants were asked to say whether the two words were related in meaning. There were pairs that were related (*train-car*) and others that were not (*train-ham*). The task was carried out only in English and the participants were highly proficient Chinese-English bilinguals who lived in Wales. Interestingly, the task of judging semantic relatedness was not the critical factor, it was just meant to mislead. The key manipulation was that in half of the word pairs, the Chinese translations of the words shown on the screen looked similar in form whereas the other half of the pairs did not. For example, in the pair *train-ham*, the corresponding Chinese translations are *huo che-huo tui*. As you will notice, these words are similar in their form and, therefore, from the researchers' point of view, they were considered formally related. On the contrary, in the pair *train-apple*, the corresponding Chinese translations are dissimilar (*huo che-pin guo*) and thus would be considered formally unrelated. But remember that in the experiment, the words were presented only in English and never in Chinese.

The authors hypothesized that, if by reading the words in English, the participants translated them automatically (and unconsciously) into Chinese (i.e. if when a language [English] is processed, the language not in use [Chinese] is also activated), then different responses would be observed for these two types of pairs. This did not happen at the behavioural level: the participants were equally fast and accurate in their responses to both types of pairs. The experiment seems to have failed. But, not so fast. During the task, the researchers also recorded the electrical activity of the participants' brains using an electroencephalogram. After analysing this signal, it was observed that the brain response was significantly different when responding to words in Chinese that were related compared to unrelated words. Remember that the task only involved stimuli in English!*

These results, among others, suggest that when bilingual speakers process one language, they cannot 'turn off' the other as if it were a light bulb. On the contrary, it would seem that both are active to a certain point during language processing. That said, how is it possible that we do not confuse and mix them? The issue of control is a little more complicated.

Without going into too much technical detail, I would like to introduce at least one of the experimental paradigms most used to understand how linguistic control works in a bilingual individual. I have chosen the paradigm of language switching because, besides having used it for more than ten years, it is to some extent easy for anyone to carry out. It is one of those experiments that you can even do at home.

One way to study the mechanisms of linguistic control among bilinguals is to explore the behavioural patterns and how their brain correlates in tasks that involve switching back and forth between languages. Consider, for example, the activity in Figure 2.3. A series

---

* For the reader familiar with this technique: the difference occurred in the temporal window where lexical-semantic effects are often detected (the *N400 component*, that is, 400 milliseconds after presenting the stimuli).

of drawings are presented to participants one after the other and they are asked to say aloud the words that the pictures represent. The drawings may appear framed with a blue or red border (the colours in particular do not matter). The participants must say the words in one or another language depending on the colour that borders each drawing. So, for example, if the subject is a Spanish-English bilingual, they may be asked to name the drawings with blue borders in Spanish and those with red borders in English. The trick is that the colour of the frame varies randomly, in such a way that there are times when two or more drawings appear in a row with the same colour border and others in which they alternate borders of different colours. For example, imagine the following sequence: car in red, umbrella in red, chair in blue, glass in blue, table in red. The correct answers would be: *car, umbrella, silla, vaso, table.*

In this sequence we find different types of stimuli, or *trials.* There

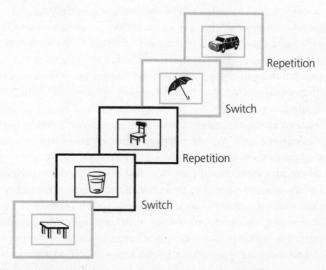

Figure 2.3. Representation of language-switching tasks. The participants have to say aloud the name of what each drawing represents. The language they should use is determined by the colour of the border. Thus, we find trials in which the language to be used is repeated (repetition trials) and trials that change (switch trials).

are trials where the language used to name the drawing is the same as the trial immediately before, as for example when *umbrella* or *vaso* appeared. We call these *repetition trials*, since the language that is used is the same. We also have trials in which the language used to name the drawing changes with respect to the trial immediately before, as when we see *silla* in blue and *table* in red. These are called *switch trials*, given that the language used to denominate them changes with respect to the one used in the previous drawing. As usual, the speed (in milliseconds) that it takes participants to name the word that describes the drawing and the error rates are measured.

I encourage you to do the experiment at home. Take six common objects, like a pair of scissors, a glass, a pencil, and so on, and put them out of sight (for example, under the table) of the person you have chosen as a participant. Now tell him that you will be showing him a series of objects and that he has to name what he sees aloud as quickly as possible. Tell him that if you show the object to him with your right hand, he will have to use his first language, and if you show him with your left hand, he will have to use his second language. Start showing objects randomly with each of your hands. If you do it at a reasonable speed, for example by presenting the new object around one second after the participant gives the answer to the previous stimulus, you can easily detect the effect we are looking for: switching languages takes longer, and you may even get a laugh when the participant makes mistakes or gets stuck. You may notice this even more so if you try this experiment on a participant whose command of the second language is not very high.

What have we observed with this task? First, that the participants are more efficient naming trials in which the language repeats (repetition trials) compared to those in which the language changes (switch trials). This is the effect and cost of language-switching tasks, that is, the time it takes to say the word in question is different in each of these two types of trials, which shows that switching from one language to another takes longer and demands a behavioural effort. Nothing surprising so far. But is this switching cost the same for a bilingual's two languages? Consider that you are doing the activity described above in his first language, Spanish, and in another in which he is rather less competent, English. Take a guess: Do you think it will be more costly for him to switch from his dominant to non-dominant language (from Spanish to English) or vice versa? Do not respond quickly, take your time . . . When I describe this experiment to my students, the vast majority guess incorrectly. The cost of language change is greater when we must switch into the dominant language (in this case, Spanish) than when switching into the non-dominant language (English). In other words, the cost of change is asymmetric: its magnitude is greater for the dominant language than for the non-dominant one (what we call 'asymmetrical language-switching costs'). The paradox, therefore, is that switching into what is easier for us is actually more costly than switching into what is more difficult for us. If you guessed it, well done – and if not, do not worry, I did not guess it right the first time either.

You now know that experimental psychologists are experts in designing experiments that yield amazing results, but what in the world does this pattern mean? Well, this asymmetry has been used repeatedly to support the idea that linguistic control of two languages is based on inhibitory processes. That is, when we want to speak in a language we have to put processes into play that reduce the activation of representations of the other language, and as such, decrease the possible interference that these representations would cause when we want to focus on the language in use. But inhibiting one language may have effects on subsequent trials in which we have to use it. That it is more costly to switch from the non-dominant language to the dominant one implies that the amount of inhibition applied to each language is different. So, if I have to say the name of what a drawing represents in my non-dominant language, I would apply a lot of inhibition to my dominant language to prevent intrusions. If then in the next trial I am asked to switch, it will be quite costly given that I will have to recover from all the inhibition applied in the previous trial. The inhibition applied to the non-dominant language would be less and, therefore, it would cost less to recover from it in subsequent trials. So, the magnitude of the switching cost would be greater when I switch to my dominant language versus to the non-dominant one. In fact, this asymmetric cost phenomenon is not unique to linguistic contexts, and also has been observed in attentional activities that do not involve language use at all. So the fact that it costs us more to return to an easier task when doing two tasks at the same time suggests that this is a property of the cognitive system, and not only of the linguistic system.

You might ask what happens with more balanced bilinguals, that is, those who master the two languages at a similar level. Suppose that the asymmetry of the magnitude of language-switching cost should be relative to the difference in inhibition applied to each language. The greater the discrepancy between the level of competition between the two languages, the greater the inhibition applied to the dominant one. Therefore, smaller differences in dominance between the languages should show less asymmetry. To make it clearer, the most balanced bilinguals should show the same amount of switching cost for both languages. In fact, that's what we found a few years back in our laboratory when balanced-bilinguals of Catalan-Spanish and Basque-Spanish participated in the task: it cost them the same to switch into one language as the other. These are true jugglers.

LINGUISTIC CONTROL IN THE BRAIN

As we have seen, some bilinguals show difficulties in language processing because of brain damage. The most common pattern is that both languages have a similar deterioration that occurs at the same time. However, there are cases in which brain damage seems to affect not so much the linguistic representations, but the voluntary control that the patient has over them. It's as if the patient is unable to focus attention on one of the languages and instead mixes them up involuntarily. Studying this verbal behaviour and its relation to damaged areas of the brain has been laying the groundwork for a better understanding of the neuronal circuits responsible for linguistic control in bilingual individuals. In fact, more and more studies are investigating how lack of linguistic control contributes to the loss of the ability to process language, something that goes well beyond examining damaged representations. In other words, the information would still be there and the problem would be how to access this information. If this is true, the case of the Formula 1 driver Fernando Alonso would exemplify such a loss of linguistic control.

Perhaps the most complete model on this subject is the one proposed by Jubin Abutalebi and David Green just over a decade ago in an article published in the *Journal of Neurolinguistics* (see Figure 2.4). They argued that different areas of the brain are involved in several aspects related to linguistic control. Of particular relevance to this skill are certain subcortical areas, such as the caudate nucleus. A deterioration in this area results in what has been called 'pathological language change' or a mixture of languages. Consider, for example, the case described by Peter Marien and his collaborators in their study of a ten-year-old boy who suffered from language problems due to a cerebral haemorrhage. The boy's first language was English, but he had learned Dutch at the age of two and a half and had been communicating with his friends and at school in Dutch. A few days after the haemorrhage, the child had problems with spontaneous language in both languages, that is, it was difficult for him to maintain a conversation. The most notable side effect was that he seemed to have lost the control of his languages and thus mixed them involuntarily.

Neuroimaging tests showed abnormal blood flow in various brain regions (what in medical terms is called 'hypoperfusion'), including in the caudate nucleus of the left hemisphere. This abnormal blood flow caused these regions to work inefficiently, which created problems for the child. Fortunately, six months later, blood flow had returned almost to normal in the frontal zones and in the left caudate nucleus, although not in other areas of the brain also related to language processing. After those six months, the child stopped unintentionally mixing English and Dutch. He still showed certain linguistic problems in both languages, especially regarding fluidity, but he no longer mixed them. The authors interpreted the relationship between the boy's symptoms and his brain damage as evidence that the frontal and subcortical areas (the caudate nucleus) are responsible for linguistic control in bilingual individuals. There are many cases of patients with brain damage in subcortical structures who show poor control of languages and we now have enough evidence, including from studies with patients suffering from Parkinson's disease, that such structures seem to be closely related to linguistic control.

These observations have laid the groundwork for the design and

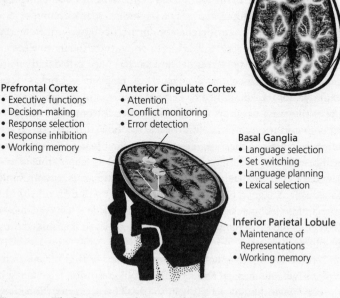

Prefrontal Cortex
- Executive functions
- Decision-making
- Response selection
- Response inhibition
- Working memory

Anterior Cingulate Cortex
- Attention
- Conflict monitoring
- Error detection

Basal Ganglia
- Language selection
- Set switching
- Language planning
- Lexical selection

Inferior Parietal Lobule
- Maintenance of Representations
- Working memory

Figure 2.4. The brain network responsible for linguistic control in bilingual speakers, using the model of Abutalebi and Green (2007).

interpretation of a good number of studies that have explored different aspects of language control using neuroimaging methods among healthy speakers. These studies have employed different types of tasks, most of which involve the need to exercise linguistic control, such as the switching task described in the previous section. Without going into too much detail, these studies show that linguistic control is exercised through the use of a brain network that involves frontal, prefrontal, and parietal areas along with the anterior cingulate gyrus and the caudate nucleus.

We also have information about what happens when we interfere with the functioning of some of these areas through the intra-operative stimulation that we described earlier in this chapter. For example, in a study conducted by Antoni Rodríguez Fornells at the

Bellvitge Biomedical Research Institute, it was observed that an interference with the normal operation of medial and inferior areas of the frontal region affected the verbal behaviour of two patients in a language-switching task similar to the one described above.

One of the central questions about the functioning of linguistic control in bilingual individuals has to do with the extent to which it involves processes and areas of the brain belonging to the domain-general executive control system. It is difficult to find a good definition of the 'domain-general executive control system', but let's put it this way: executive functions are those that we put into play when we want to do something without getting distracted. It's a little more complicated than that, but for the time being this definition suffices. These control processes are triggered continuously and allow us to maintain the goals we want to keep active in our mind, as well as ignore stimuli or information that can interfere with the appropriate behaviour to achieve them. If you have seen the movie *Finding Nemo*, you may remember Dory, the blue fish that accompanies Nemo's father on his search but continuously gets distracted and confused. Dory lacks some important parts of the executive control system like working memory.

In the case of linguistic control, the goal is to speak in the desired language and the distracting information is the language that is not being used. Given this parallelism, it is reasonable to think that linguistic control processes use the resources of the domain-general executive control system. However, the behavioural results that we have at present from neuroimaging studies indicate that although there is some overlap, it is only partial. We will return to this question and discuss it in more detail in Chapter 4.

FORGETTING THE FIRST LANGUAGE

Most studies on bilingualism seek to understand the acquisition processes and use of a second language. To put it another way, scientists, and I dare say most people, are interested in understanding how one goes from being monolingual to bilingual or how one grows up being bilingual. The latter makes sense, because it's quite common. However, some researchers have asked a question that is somewhat related to the previous one and can tell us a lot about how we learn . . . and unlearn: What happens when one language replaces another?

This is a topic that has to do with what we call 'first-language attrition'. There are numerous works that have explored how the acquisition of a second language affects the use of the already established first language. The patterns of interaction between the two languages are complex at all linguistic levels. In many cases one language is not replaced by another, but peculiarities can be seen in the use of the dominant language.

I had the opportunity to observe these interactions first-hand when I lived in Boston and was conducting experiments on bilingualism in the Cognitive Neuropsychology Laboratory directed by my mentor, Alfonso Caramazza at Harvard University. In addition to posting ads all over campus to recruit bilingual Spanish-English participants for my tests, I was also looking for participants informally at the many parties organized by my Latino friends. You know, life isn't just about science. Between margaritas, mojitos, and too much Salsa music for my taste, I more or less managed to explain the type of studies that I was carrying out. My goal was clear: I had to get their emails or phone numbers to contact them a couple of days later, when their attentional abilities would be more in tune. The important thing was to get the contact. As expected, when I called them the following Monday with the intention of setting up an appointment to perform the experiment, many of them reacted with surprise and claimed to have no idea what I was talking about, and in many cases didn't remember who I was (some even refused to remember having been at the party at all, but that's another story). Although I have to admit that this strategy for recruiting participants was unconventional, it worked and I managed to carry out my post-doctoral experiments.

I tell you this not because it was strange to meet young people who, even though their first language was Spanish, had a clear dominance in English, but rather because they were enrolled in courses in Spanish as a second language, that is, courses for native speakers of English. When I was talking to them, I could notice the effect that English had on their Spanish, both at the grammatical and lexical levels and even at the phonological level. The two languages were interacting in such a way that one was 'eating' the other. Several reminded me of some Catalan speakers who immigrated to Mexico when they were young as a consequence of the Civil War, and who spoke in Catalan with Mexican prosody, a very curious and endearing thing. This effect of the learning of one language over an already established one shows the plasticity and dynamic nature of our brain.

The number of children who are adopted by caregivers who speak a different language is considerable. Without thinking too hard, I can name about ten acquaintances who have adopted children from Russia, China, Vietnam, or Ethiopia, and none of these acquaintances knew (or know) the languages spoken in these countries. In many of these cases, the children no longer have contact with their first language and become immersed in a second (or third) language. There is no doubt that this situation entails a loss of skills in the dominant language, but are there any traces left in the brain of the first language once they reach adulthood? Or is the cerebral plasticity such that these adoptees will completely forget what used to be their first language for some months, and in some cases for a few years? Can the brain forget a language?

These studies are difficult to carry out, and perhaps because of this, there are only a few of them. In one of these studies directed by Christophe Pallier at the Institut National de la Santé et de la Recherche Médicale (National Institute of Health and Medical Research) in Paris, eight Korean adults were selected who had been adopted by French-speaking parents. The age of adoption varied from three to eight years of age, which meant that these children had already acquired Korean when they left their native country. Nevertheless, they all claimed to have completely forgotten their mother tongue and to have had no problems learning and using French. The authors asked these participants to perform several tasks in which Korean came into play. For example, they heard a series of recorded phrases in several languages that were typically unfamiliar to French speakers (e.g. Japanese, Korean, Polish). The participants were asked to say whether they believed that each of these phrases belonged to Korean. In another one of the exercises they were shown a written word in French followed by a recording of two words in Korean. The participants had to decide which of them corresponded to the translation of the French word. The performance of the adoptees in these activities was compared with that of another group whose mother tongue was French but who had no experience with Korean; that is, a kind of control group. The hypothesis was clear: if the adoptees maintained some kind of knowledge of their mother tongue (Korean), unconscious or indirect as it may be, their accuracy would be greater than that of the other group.

The results did not confirm this hypothesis; in fact, the accuracy was identical for both groups. Korean had disappeared from their mind, even among those who had had quite a bit of experience with and exposure to the language (i.e. eight years). The authors went a step further and decided to analyse the brain activity of the two groups during a task involving Korean. After all, even though their behavioural performance did not show traces of the lost language, maybe their brain activity would. In this task, the participants heard a series of recorded phrases while their brain activity was recorded. When analysing the brain activity of the French participants (the control group) while listening to French or Korean, there was greater activity in the classic areas related to language when the sentences were reproduced in French. This makes sense because these individuals had no previous contact with Korean. How did the brain respond among the adoptees who previously did have contact with the Korean? It was exactly like the French participants. That is, the brain of those adults who had grown up with Korean for several years and those who had grown up without Korean reacted in the same way. It was as if none of them had ever learned it. The group of adoptees had forgotten their mother tongue.

However, another study conducted by Jeffrey Bowers at the University of Bristol yielded a surprising result. The study explored the ability of adults whose mother tongue was English to learn a phonological contrast that exists in Zulu and Hindi but not in English. Remember that in Chapter 1 we talked about how our ability to discriminate between sounds that we are not exposed to in our environment diminished by the age of one. Some of the adults had had contact with the two languages during childhood, but at the time of the experiment they claimed to have lost all knowledge of them. A control group was composed of native English speakers who had no experience with Zulu or Hindi. The question was whether those who

had used those languages during childhood could 'relearn' the phonological contrast faster than those in the control group, which
would suggest that there was still a trace of that language in their
brain. The results of the study were clear. At the beginning of the
sessions, the two groups showed equally poor performance and it
was difficult for them to differentiate the sounds. This reaffirmed the
idea that the group of participants who had been exposed to those
languages had lost all knowledge of them. However, as the test
advanced, the group with previous experience with Zulu or Hindi
was able to discriminate between the sounds more efficiently than
the control group. These results suggest that the group with previous
experience with Zulu or Hindi had maintained some knowledge, at
the phonological level in this case, of a language they had stopped
using many years ago. Their brain had saved some of that experience
from childhood even though they were not aware of it.

Considering these results, it is premature to conclude that a language can be completely forgotten after it is no longer used. But these
studies are important because not only do they provide us with information about the interaction between languages, but also about
brain plasticity, and even how we forget language.

We have come a long way in this field, but we still have much to
discover. If we could analyse the mechanisms that the most wonderful animal in evolution, the Babel fish, uses when it translates all
languages, this task would be much easier. Unfortunately, Douglas
Adams took this secret with him, although apparently engineers are
trying to decode it.*

* An example of this is the company Waverly Labs, which aims to design devices
that allow communication between people of different languages through simultaneous translation. One of its products is a set of two small headphones, one for
each individual involved in a conversation. Each message that is emitted is cap-
tured by the telephone to which the headphones are connected and the message is
translated into the desired language for both people. So, two people who speak
different languages can hold conversations. It's as if each one of the interlocutors
had a Babel fish in their ear.

3

How Does Bilingualism Sculpt the Brain?

In many parts of the world bilingualism has an inevitable sociological and political dimension because it is often linked to other factors such as emigration and national identity. This leads to interested, and not entirely objective, claims about the dangers or advantages that the bilingual experience can bring. Some say 'bilingualism causes problems for linguistic development and use' or, more extreme yet, some of the great minds of a few decades ago assumed that bilingualism could result in mental illnesses such as schizophrenia. Although today such exaggerated views are not all that frequent, there are still some who warn about the damage that bilingualism can cause. This kind of claim is often used to question models of bilingual education.

On the other hand, some recent studies, which seem to indicate a more efficient development of certain cognitive abilities associated with the use of two languages, have been publicized by the media as evidence that bilingual speakers are more intelligent. This is not an entirely new opinion either, as we have seen in the previous chapter: in the 1960s, the renowned neurosurgeon Wilder Penfield asserted in an interview published in a Canadian newspaper that the bilingual brain was superior. Fifty years later, I was interviewed for a piece for *The New York Times* that made a very strong claim: *Why are bilinguals smarter?* Again, social and political agents that promote national identity in places where two languages coexist use this to promote bilingual education. In Spain, I constantly witness this polarization and the use of bilingual studies as a weapon when I do media interviews in which there is interest in highlighting one or the other aspect, but not so often both.

The scientific question that interests us here is the effect of bilingual experience on language processing, cognition, and brain development. In this chapter, we will focus primarily on the former, and leave the issue of how the bilingual experience influences other cognitive domains for the next chapter. To analyse the effect of bilingualism on language processing, it is necessary to compare the performance of bilinguals with that of monolinguals and, like any comparison between groups of individuals (different social strata, sexes, countries, and so on), the conclusions drawn are always ... delicate. To put it another way, it is not politically correct to discover that women are better at a particular intellectual activity than men, or vice versa.

To avoid confusion, let's start by stating a truism: the bilingual experience does not seem to have dramatic effects on the linguistic capacity or any other cognitive domain of individuals. We all know bilingual speakers who express themselves without apparent difficulties in their native language (and in the non-native language) or who, at least, do not seem to find it any more complicated than monolingual speakers. So we can state that acquiring a second language does not seem to have devastating effects on the use of the first one, unless, as we have seen in the case of the adopted children mentioned in Chapter 2, it is no longer used. On the other hand, bilingual speakers do not appear to be 'smarter' than monolinguals, and there seems to be no remarkable difference between their cognitive abilities. So don't worry about whether your opponent in a chess match is bilingual or not. Having stated the obvious, next we will look at several studies that show certain differences between bilinguals and monolinguals in some cognitive capacities. What is interesting about these differences is that they are useful for understanding how varied cognitive processes interact with each other. Let's start with language and answer the question of whether the bilingual experience involves some kind of difficulty in linguistic processing.

INTERFERENCE

I usually give this example to my students: Juan and David are going to play a tennis match. Juan practises tennis every afternoon for three hours, while David does so for only one and a half hours and the rest of the time he plays squash. Who do you think will win the game? The majority of the students, showing admirable intelligence and prudence, affirm that they do not possess enough information and that, surely, there are many other factors that they would have to know in order to make a sensible prediction. But I don't let them off the hook that easily, and I give them more pieces of information: Juan and David are identical in all other aspects related to tennis: they learned to play at the same age, they are equally tall, and have the same motor coordination, etc. At this point, the students bet on Juan, reasoning that he practises twice as many hours as David and that, all other things being equal, Juan should win. It's true that they also say that David can play two sports and Juan only one, but that's another story.

Chances are you have already noticed the analogy between practising a sport and practising a language. Juan practises a single sport (tennis) every afternoon, that is, one language (Spanish), while David practises two sports (tennis and squash), that is, two languages (Spanish and English). Juan is monolingual and David is bilingual. Therefore, if the analogy were valid, it would be expected that the greater frequency with which monolinguals practise their only language, compared to bilinguals, will result in differences in the efficiency with which they use it. After all, we know that the frequency with which, for example, we use words affects the reliability and speed with which we retrieve them during the speech production and we recognize them during their comprehension. Speakers tend to retrieve with more speed and accuracy words that are frequent (*table*) compared to others less frequent (*cavern*). In addition, we tend to fall into 'tip-of-the-tongue' situations when trying to retrieve low-frequency words (this would never happen to anyone with the name of their mother). How do we know this? I will try to prove it to you. Can you tell me the name of the mythological creature that is half-man and half-horse? Tic-toc, tic-toc, tic-toc. If you came up with the name, congratulations, you can continue reading peacefully; if you have it on the tip of your tongue, let me be a little ornery and not give you the answer until the end of this section. Well OK, I will give you a clue: it starts with 'ce'.

There are several studies that have shown that certain linguistic abilities are affected by bilingualism. Bilingual speakers have a slower and less reliable access to the lexicon than monolinguals in speech production tasks. We know this thanks to experiments that have used the technique of naming what is presented in drawings: participants are simply asked to say aloud what appears on a computer screen as quickly as possible and try not to make mistakes. How long do you think it takes to start articulating the name of a picture from the time it first appears on the screen? Young speakers are able to perform this task in 600 milliseconds on average. Not bad, right? Especially if we consider that they are choosing the desired word among the thousands stored in their mental lexicon.

Bilingual speakers perform this task more slowly and with more errors than monolingual speakers, as can be seen in Figure 3.1. This would not be too surprising if this happened when we compared bilinguals naming drawings in their second language with monolinguals naming drawings in their only language, since it would not be entirely fair. After all, the fact that bilinguals were less efficient in their second language is not surprising, since we often find differences among the bilinguals themselves with respect to their performance in the first and second language. In addition, we also know from other studies that there is a negative correlation between the age at which words are learned and the speed and accuracy with which they are processed; for earlier ages of acquisition, this is faster and more accurate. What is more surprising is that the difference in efficiency between bilinguals and monolinguals is also observed when they both name drawings in their first language (the only one in the case of monolinguals). This happens even for highly proficient bilinguals. It is true that the difference between one and the other is not very large (around 30 milliseconds), but the picture-naming exercise is relatively easy. We do not yet know how a difference of this size can be magnified (or reduced) by evaluating the verbal behaviour of speakers in more complex linguistic situations.

To be fair and more precise, it can be said that these differences arise to a greater extent for words that do not resemble their translations in the other language (*table* in English; *mesa* [*table*] in Spanish); they are what we call 'non-cognate words'. However, words that are similar (*guitar* in English; *guitarra* in Spanish) are not as susceptible to the slowdown associated with bilingualism.

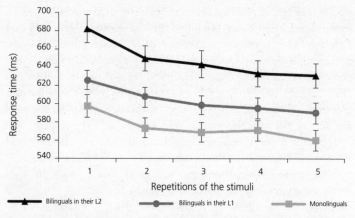

Figure 3.1. Results of bilingual and monolingual speakers naming drawings in the first and second languages. On the vertical axis, the response time is represented in milliseconds: the slower the time, the higher the line is. On the horizontal axis, the various repetitions of the stimuli are represented. As they repeat the task, the response time decreases. However, the difference between the three conditions remains constant.

Other proof that access to the lexicon is less efficient in bilingual speakers comes from the observation that they tend to fall into tip-of-the-tongue states more often than monolinguals. As you can imagine, these studies are complex, because they are difficult to orchestrate. Tamar Gollan from the University of California, San Diego, has found a way to do so by presenting a series of definitions of low-frequency words and asking the participants to say the corresponding term aloud; basically the same thing I did earlier with the name of the mythological animal. In addition, and surprisingly, this tip-of-the-tongue state happens even when bilinguals are allowed to say the corresponding word in either of their two languages. That is, it does not seem that this difference is due only to one of the languages blocking access to the other.

Some of the activities that are very often used to evaluate the linguistic capacities of a patient with brain damage are those involving verbal fluency. The activity is very simple and you can practise it with anyone (I think there was once a TV game show that did it). Here are the instructions: 'Please name as many animals as you can in a single language in a minute and without repeating any word.' This task requires quick access to the lexicon, as well as control over what has already been said to avoid repetition. It has also been shown that bilingual speakers list fewer examples than monolingual speakers, which would suggest that their access to words is more costly.

These results, among others, would suggest that the bilingual experience affects the efficiency with which lexical access processes work. These effects can be due to the differences in the frequency of use of each language, or also, in some cases, to the interference that a second language can cause during first-language processing. That interference, as we have discussed in the previous chapter, results from the fact that bilingual speakers cannot turn off the language that is not in use. Just look at, for example, the verbal fluency exercise that I have just described: we have to avoid producing words in another language, and to do so the bilingual speaker has to continually block the possible interference that these words could create. Therefore, under a situation of temporary pressure, in which we have to say as many words belonging to a specific category as we can within a limited period of time, it is possible that this interference results in a poorer performance.

There are other examples of this in which it is evident that the use of a language can have negative effects on the recovery of the other language's representations later on. Imagine that we asked a group of bilinguals to name a series of drawings in their second language. After that, we ask them to name those same drawings in addition to new ones, but now in the first language. In principle, one might think that in this second task the reaction would be faster for the drawings that had already appeared, since, at least, it would be easier to recognize them. Well, as it turns out, it is not. The activity seems to be more costly for the drawings that appeared before than for those that were newly added. It is as if having named something in a second language makes it difficult for us to do so in the first, which would suggest the appearance of interference between them or, if you will, how costly it is to recover from the inhibition exerted during second-language production. We mentioned a similar finding when we talked about asymmetric costs associated with language switching in Chapter 2.

Let us now consider the effects of the tip-of-the-tongue state. As we have seen, when we fall into a state like this, it is usually in situations in which we try to retrieve a word that we do not use often or is of low frequency. It is reasonable to think that the regularity with which the bilingual uses the words in each of their languages is less than that of the monolingual. To put it more simply: all the time I spend using English, I am not using Spanish. Therefore, we could say that for a bilingual, there are more words of low frequency than for a monolingual, and since those are the ones that can make us fall into a tip-of-the-tongue state, a bilingual is more likely to suffer from this phenomenon in both languages.

It must be noted, however, that the magnitude of the effects described above is not dramatic and that there is much variability within each group of speakers. Let's put it another way: we cannot make good predictions about the verbal behaviour of an individual based only on their bilingual status, since there are many other variables that will affect their linguistic performance. The effect that bilingualism can have on linguistic efficiency is only one factor, but there are many more. Returning to the analogy between practising tennis and linguistic competence: my students were smart and cautious in saying that they lacked information when they only knew the number of hours Juan and David practised, and that, therefore, they could not guess who would win the tennis match. Furthermore, we should be cautious when making statements about the linguistic capacities of specific people, be they bilingual or monolingual. And now I will fulfil my promise and relieve those of you who have fallen into a tip-of-the-tongue state from the question of naming the half-man, half-horse mythological being. It is a centaur.

OUR MENTAL DICTIONARY

Another effect that seems to be associated with being bilingual has to do with a possible reduction in vocabulary size. Do bilinguals really know fewer words than monolinguals? Let's walk through this carefully and start from the beginning.

The ability to learn new words remains open throughout life and, in fact, we never stop doing it. That is, like other language-related skills, such as the acquisition of new sounds, that are dramatically reduced with age (remember the phenomenon of perceptual adaptation discussed in Chapter 1), ageing does not seem to affect learning new lexical items too much. Think about the words added to the Oxford Dictionary in June 2018: antwacky, beerfest, binge-watch, cromulent, Facebook, first-wave, gabster, heteroglossia, hip-pop, impostor syndrome, indica, lab rat, pansexual, piffy, sativa, screed, scrid, scrim, scrum-down, silent generation, spad, twine, ungendered, unween, walkative. Given that it is difficult for dictionaries to add new words unless speakers use them on a daily basis, you may know some of them, but you may also have only learned them recently. I must confess that I did not know several of them, such as *impostor syndrome* and *gabster*, so today I have already learned two. The number of new words that we are acquiring depends on the linguistic richness to which we expose ourselves. In other words, it is difficult to learn new words when language experiences are reduced to sports articles and talk-shows on TV; other types of activities are more stimulating and linguistically – and cognitively – challenging, and I kid you not.

Knowing that this ability is present throughout life, have you ever wondered how many words you know? Two thousand, 10,000, 20,000 ... well actually it's quite a bit more. According to some calculations, a speaker with higher education usually knows about 35,000 words. That's not bad, right? Obviously, this does not mean that we use most of them on a regular basis; in fact, we only use around 1,000 words on a daily basis. (Don't be disappointed: Cervantes is said to have used a total of about 8,000 words in all his works.)

Let's pause for a moment and analyse a recent study that estimated the size of the vocabulary of Spanish speakers, which I think wonderfully exemplifies how we can use new technologies to answer interesting questions. That was the purpose of a study that involved my colleagues Jon Andoni Duñabeitia and Manuel Carreiras at the Basque Centre on Cognition, Brain, and Language (BCBL). Taking advantage of the fact that the vast majority of us have a cellphone, tablet, or computer with an internet connection, the researchers launched a platform where in just four minutes a good estimate of the user's vocabulary level could be elicited. You can check it by searching for the vocabulary test put together by Ghent Univesity. The task proposed is quite simple: a series of strings of letters appear on the screen and the participant must indicate whether each corresponds to a real word in the language being tested or not, a test which we call a *lexical decision task*. It seems easy, but hold on for a second. We all know that *home* is a word and that *hofe* is not. But what about *abstemious*, *ocubavious*, *aplomb*, and *oclomp*? I will not make it that easy by just giving you the answer, for the time being.

One of the virtues of this task is undoubtedly the agility with which it can be completed, since every time a user starts it, the system randomly creates 100 strings of letters drawn from a database of almost 50,000 real words and invented words. If we add to this random sampling the easiness of performing the test on devices that we handle every day, within just a few weeks of launching there can be hundreds of thousands of people taking it. Knowing the percentage of success of each of the participants, we can estimate the average size of the vocabulary of a speaker type, let's say a speaker of Spanish (the way the index is calculated is a little more complicated, but this is sufficient for our purposes here). As seen in Figure 3.2, the number of words we know increases as we get older (so it might not be a fair match for you to play *Words with Friends* or *Scrabble* with your grandfather). By the way, *abstemious* (referring to moderation in eating or drinking) and *aplomb* (meaning self-confidence) are real words; *ocubavious* and *oclomp* are not.

Now, there are two considerations that we must make before moving on to discussing the effect of bilingualism on vocabulary development. First, there is always time to learn new words and, in fact, we do this continuously, even if we do not realize it. Second, the richness of our vocabulary is related, to a large extent, to the exposure we have to contexts in which the use of new words is more frequent.

Several studies have shown that bilingual individuals have a smaller vocabulary in their two languages than monolinguals. Consider, for example, the studies of Ellen Bialystok and colleagues at York University in Toronto. In one of these studies, the receptive

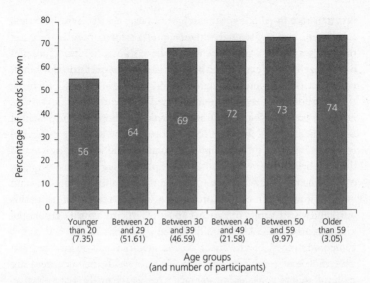

Figure 3.2. Percentage of words known to participants according to age ranges. As you can see, the percentage increases with age. The number of participants for each age group is presented in parentheses.

vocabulary of almost 2,000 children between the ages of three and ten was explored. *Receptive vocabulary* refers to words that are recognized when we hear them and from which we can identify the meaning, regardless of whether we use the word habitually. To carry out this experiment the researchers used a standardized test for different ages called the Peabody Picture Vocabulary Test, which they administrated to monolingual children of English and bilingual children of English and various languages. The vocabulary score was higher for monolingual children of all ages. Interestingly, the kind of words in which monolinguals tended to outnumber bilinguals were used mostly in domestic contexts. And when the vocabulary that is used in school was evaluated, the difference between the two groups disappeared. It makes sense, right? After all, in the school context all children were exposed to the same words (at least in this study). This last detail is important, since the size of the school vocabulary is a good predictor of academic performance. The fact that there was no difference in school vocabulary would suggest that bilingual children would not be affected in their school performance. In any case, this study and other subsequent work have shown that the reduction of the size of the lexicon for bilinguals extends on into adulthood, from twenty to as much as eighty years.

These findings have to be interpreted with care and, as you can imagine, they are the perfect ammunition for those who are against bilingual education. First, we have to look at how large the difference is in vocabulary for bilinguals and monolinguals, something we call the *magnitude* or *effect size*. Let me explain: imagine that we take a cold medicine that, according to proven studies, shortens the duration of the symptoms. In other words, when we randomly administer this medicine to one group of patients and a placebo to another group, the symptoms of the cold generally disappear earlier in the first group compared to the second. Perfect; we're convinced, let's go buy the medicine. But wait a minute. Ask yourself how much the symptoms are reduced, that is, consider not so much whether the medicine is effective, but *just how* effective. If it turns out that the symptoms will last two days less, you may want to buy the medicine, but if they will only last two hours less, you may want to think again (after all, you will have a cold for almost the same time with or without the medicine). With vocabulary reduction, the same thing happens: the result of the test from the aforementioned study has an average of 100 and a standard deviation of 15. This basically means that the majority of children score between 85 and 115. But what is the average of bilingual children? Between 95 and 100. And that of monolinguals? Between 103 and 110. That is to say, all are very close to the average of the general population. Therefore, it is true that there is a reduction in the size of the vocabulary associated with bilingualism, but this is relatively modest.

On the other hand, we may be tempted to apply the group norm to particular individuals and think, for example, that if our child grows up in a bilingual environment, his vocabulary will necessarily be smaller than if he grows up in a monolingual environment. Stop right there; applying a group norm to an individual is inappropriate and, in this specific case, even less so. Let's look at Figure 3.3, which shows the distribution of scores for bilingual and monolingual children on a vocabulary test. On the horizontal axis we find the test

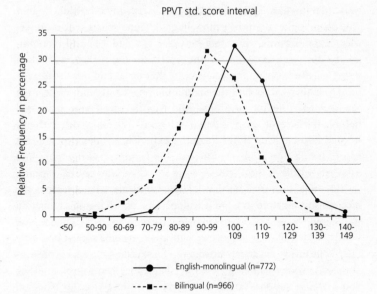

Figure 3.3. Distribution of vocabulary scores for bilingual and monolingual speakers.

scores, and on the vertical axis we see the percentage of children who obtained those scores.

The higher the point of each line, the higher percentage of children received the range of vocabulary scores shown on the horizontal axis. For example, we see that there are about 7 per cent of bilingual children who scored between 70 and 79, and that there are only about 1 per cent of monolingual children in that range. We can also observe that the majority of monolingual children scored between 100 and 120, while most bilinguals scored between 90 and 110. Therefore, the average score for each group is different, being higher for monolinguals. In other words, monolinguals generally know more words. But we already knew that. However, what is striking about this graph is that there is a great overlap in the distribution of scores between the two lines, that is, between the scores of monolingual children and those of bilinguals. This means that there are many bilingual children who score higher than their monolingual counterparts. For example, there are bilinguals who score between 110 and 119 and monolinguals who score between 90 and 99. If we took, then, a random bilingual child, say, my son, it is clear that he should not necessarily have a smaller vocabulary than a monolingual child nor, in fact, than the average monolingual group. And why is that? As we have said before, the size of the vocabulary depends on many other things beyond bilingualism. If our linguistic experience is more focused on sports articles and talk-shows than on *National Geographic* and scholarly works, it is difficult to write like Cervantes or Shakespeare.

But what if it turns out that bilingualism leads to problems in the mechanisms involved in word learning? That is, what if this reduction in vocabulary was not due to the frequency with which the children use words in each of their languages, but rather to some sort of linguistic interference that negatively affects the formation of lexical represent-ations? As we have already noted in Chapter 1 when we were talking about babies, this does not seem to be the case, since basically bilingual individuals know more words than monolinguals if we add together all words in both languages. It makes sense, because for many words the bilingual will also know their translation equivalents, regardless of how different they may be. It seems that bilingualism does not interfere with the formation of lexical representations and, therefore, with the acquisition of words. Most likely, the reduction in the vocabulary asso-ciated with bilingualism has more to do with the frequency of use and the likelihood of exposure. The greater these two factors are, the more likely we are to encounter new words that we must learn. It is reason-able to think that, all other variables being constant, bilinguals are less exposed to each of their languages than monolinguals and, therefore, less likely to encounter low-frequency words. What is not used tends to be either not learned or forgotten. But let me be clear that bilingualism is only one of the variables that can affect vocabulary size and even so, it is likely not the most relevant.

Before moving on to the next section, I would like to highlight one of the practical consequences of these studies. Many tests looking at linguistic development among children and linguistic assessment for patients with brain damage are standardized considering the verbal behaviour of monolingual speakers. That is, the standard with which we compare the performance of a particular person comes from monolingual speakers. Comparing and contrasting a bilingual's capacity to this referential group can lead to confusion and mis-diagnosis, since the comparison is not even adequate when evaluating the vocabulary of the bilingual in their first language. So do not worry too much if your bilingual children do not score extremely well on a vocabulary test; maybe they are not having learning problems, but are being measured with the wrong scale. In fact, it is possible that they are learning more words than other monolingual children but, of course, in two different languages.

I believe that up to now I have fulfilled my promise to not give advice, but allow me some liberty here: if you really care about the development of your children's vocabulary, then expose them to a rich, stimulating, and challenging linguistic environment. As the pedagogue and writer Amos Bronson Alcott said, 'a good book [is one] which is opened with expectation, and closed with delight and profit'. Do not worry about which language you read to them in; but if it is in both, all the better.

BILINGUALISM AS A SPRINGBOARD FOR LEARNING OTHER LANGUAGES

Perhaps you have heard before that people who speak two languages have an easier time acquiring a new one. Is this another urban legend? Given my not-so-strong ability to learn languages, I have always been intrigued by this assertion, which, in my view, has an interesting and trivial side. The trivial view is that if a bilingual speaker is faced with a new language that is in some aspects similar to one he already knows, it might be easier to acquire those similar aspects. I lived for a year in Trieste and, although I never received formal classes, I was able to understand a good number of words in Italian. It was evident that from my knowledge of Spanish and Catalan, learning Italian would be relatively easy for me, and I say this because, as I said, my skills in this regard are quite modest. But of course, most of the words were famil-iar to me: if I came across one that was not similar to Spanish (*donna* and *tavola* in Italian are *mujer* and *mesa*, respectively, in Spanish), it was very possible that it was similar in Catalan (*donna* and *tavola* in Italian are *dona* and *taula*, respectively, in Catalan), and vice versa. Italian shares many cognate words with Spanish and/or Catalan; that is, words that have a common origin and maintain a formal similarity. It's true that some other Italian words were not similar to any of my languages (e.g. *quindi* in Italian means something like *por tanto* [*therefore*] in Spanish). There were also false cognates, or very similar words that do not mean the same (*gamba* in Italian means *leg*, but in Spanish *gamba* means *shrimp*), but that is another story. In any case, my knowledge of Spanish and Catalan – two languages that were similar to the new one (Italian) – obviously gave me a greater advantage than if I had only known one, either one of them; that is, if I had been monolingual. Notice that here I have focused on word similarity between several languages, but the same argument, or one even more substantiated, can be extended to the acquisition of the phonological repertoire of a new language or its grammatical properties (remember, for example, the problems that English speakers have when learning the grammatical gender of Spanish words). That is, the similarity between languages can help to transfer certain properties from those we know to new ones. Although this can sometimes lead to a certain confusion, in many cases it favours learning. This confusion is often found at times such as when we are faced with false cognates (*terrific* in English has nothing to do with *terrorífico* [*terrifying*] in Spanish) or when we do things like transfer the grammatical gender from the words of one language to another (*sonne* [*sun*] in German is feminine but *sol* [*sun*] in Spanish is masculine; *mond* [*moon*] in German is masculine but *luna* [*moon*] in Spanish is feminine). In any case, the most interesting aspect derived from the question of whether the knowledge of two languages can favour the learning of a third language has to be separated from the extent to which it comes from the similarities between the languages in question. This is the trivial view; let's now go to the more interesting part of the claim.

Some studies have shown that adult bilingual speakers are better than monolingual speakers at acquiring words from a new *invented* language. In one of these studies, led by Viorica Marian at Northwestern University, researchers taught words from an invented language to three groups of participants: English-Mandarin bilinguals, Spanish-English bilinguals, and English monolinguals, by presenting the invented words paired with their English translations. For example, they had to learn that *cofu* meant *dog* in the new language. Why an invented language? Because this way the researchers could ensure that the similarity between the new words and those in English, Mandarin, and Spanish was minimal. That is, the possible transfer between the properties of the languages of origin and the new language could be controlled. The results showed that both groups of bilinguals were able to learn more words than monolinguals and, in addition, this advantage was maintained at least one week after the learning session. We still have to investigate the mechanism that allows for this advantage. This will help us to know to what extent this occurs in all types of bilinguals or only those who learned their two languages in childhood, as was the case in the study presented here. In any event, what we know so far is that it seems likely that the knowledge of two languages helps to develop certain mechanisms that are put into play during the acquisition of words from a subsequent language.

Similar observations have been obtained in contexts outside the laboratory, such as the English school performance of bilingual and monolingual children in the writing and the morphological and orthographic knowledge of that foreign language.

We have not yet explored an area where I think it is very likely that we will find differences between bilinguals and monolinguals with respect to the acquisition of a new language: linguistic control. As we saw in Chapter 2, acquiring a second language and being able to use it requires learning how to control it. In this sense, when a bilingual and a monolingual face the acquisition of a new language, it is reasonable to think that the former has already developed some control processes that they can apply or transfer. Let me make the following analogy: when faced with a new language, the bilingual has to learn to juggle with three balls, already knowing how to do so with two, while the monolingual has to learn from the beginning. It is reasonable to think that the bilingual may have an advantage in this case. In fact, some results from the language-switching paradigm presented in the previous chapter would suggest that this is the case.

Remember that switching to the dominant language is more costly than switching to the non-dominant one, although this applies when there is a clear difference between the mastery of languages. In speakers with a good command of both languages, no such asymmetry is observed, and the switching cost is equal for both, which is, to a certain extent, logical. Thus, if we ask a highly proficient bilingual to carry out a language-switching task according to the colour of the border in which the drawings appear, in one of their dominant languages and in a third language that they do not know as well, we should find new asymmetrical switching for these languages. As it turns out, this is not the case: the pattern of switching cost between the dominant language and a third language is exactly the same as when the bilingual performs the task in his or her two dominant languages. It is as if he or she were applying the same linguistic control mechanisms regardless of language dominance. And this could give an advantage in the use of a third language – not so much in learning it, but in how to use it and control it, which will be reflected in the fluidity with which it is spoken.

EGOCENTRISM AND THE PERSPECTIVE OF THE 'OTHER'

Do you remember the last time you asked someone on the street how to get to a certain place? This is a response that may be familiar: 'Cross this street here, then turn right, when you find the second roundabout, take the third exit, and then turn right on the second street and you have arrived!' Do you ever have the feeling that it would be better to not have asked at all, as expressed in the cartoon below (Figure 3.4)? When someone gives us directions of this kind, the speaker has in his mind a map of the route that we should take. So he has the advantage of being familiar with all the places that you have to pass by. But for you, the question is more complicated, since you lack such a map in your mind and you have to build it according to how the other person describes. A small mistake in your mental map such as a turn to the right instead of the left, and that's it, you are lost.

PANEL A

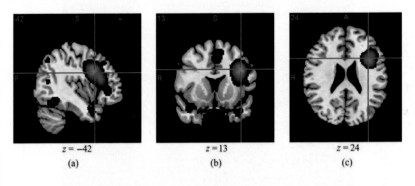

| $z = -42$ | $z = 13$ | $z = 24$ |
| (a) | (b) | (c) |

PANEL B

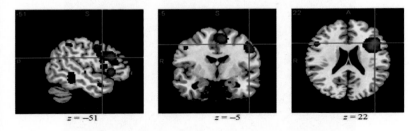

| $z = -51$ | $z = -5$ | $z = 22$ |

Plate 1 In Panels A and B, the result of the meta-analysis can be seen for the highly proficient and moderate- to less-proficient bilinguals, respectively. Note that, by convention, the right part of each brain image corresponds to the left hemisphere.

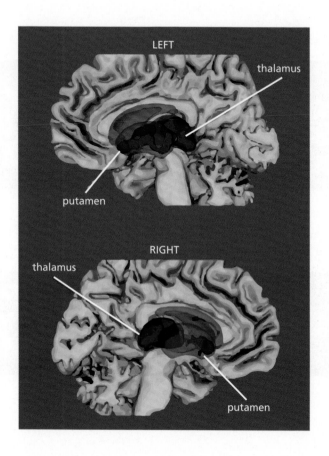

Plate 2 In the cross-section, the structures corresponding to the basal ganglia and thalamus are observed in red and the areas in which the bilinguals show increased volume compared to their monolingual counterparts are shown in blue.

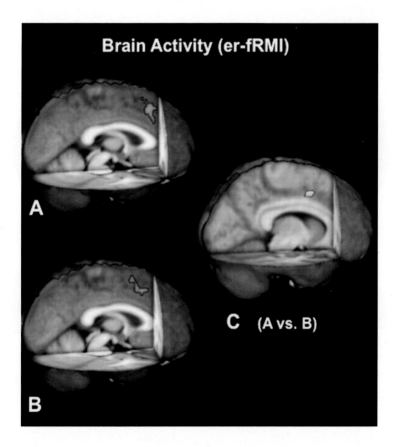

Plate 3 Neurofunctional differences between bilinguals and monolinguals in the flanker task. How the anterior cingulate cortex is activated during conflict resolution is shown in panels A and B, for bilinguals and monolinguals respectively. As can be seen, the activation for monolinguals is greater than for bilinguals. In panel C, the calculated difference between the observed activation in monolinguals and bilinguals is presented.

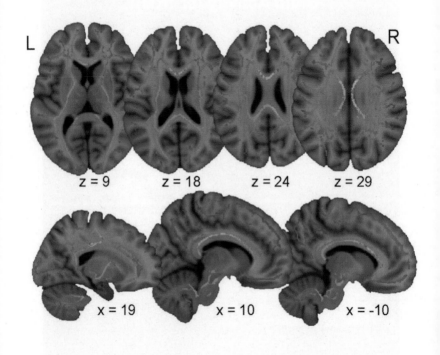

Plate 4 Differences between bilinguals and monolinguals on a measure of white matter called 'fractional anisotropy'. The coloured areas in red correspond to those in which there is a significant difference between the bilingual and monolingual groups. These areas are located in the corpus callosum and extend to the superior longitudinal fasciculus and the inferior frontal-occipital fasciculus.

Figure 3.4

This cartoon exemplifies how difficult it sometimes is to have clear communication, partly because the perspective of the one giving the directions is different from the one who receives them. When we communicate with someone, it is essential to know the perspective that the listener has of the context. We must put ourselves in the place of the other, and try to guess what they know about the subject we are talking about, and to what extent our point of reference is common. Otherwise, communication becomes very difficult. Think, for example, how many times mistakes are made when an appointment is made with someone who is in another time zone. You set a time to talk at six. But is this six o'clock for our interlocutor who is in London or for the other person in Madrid who is an hour ahead? What is the point of reference, ours or the other person's? We have to establish a common point; if not, we will be lost again. When we hold a conversation, it is as if we were dancing with someone. It is a collaborative activity in which our movements depend on and continually combine with what the other does. Interlocutors do the same when they talk. Well, if they want to be understood.

Being able to put ourselves in the other's place is difficult, and in fact we often do so with what is called 'egocentric bias', or the tendency to think that the other person has the same information and perspective as we do about a specific situation. Basically, if I see it one way, I think that you do too. As it turns out, the bilingual experience seems to help develop the ability to put oneself in someone else's shoes. Let's look at a study carried out by Katherine Kinzler and Boaz Keysar at the University of Chicago, because it will serve as an example of how to study a speaker's perspective. The experiment is simple and ingenious.

Two people participate in the study. One is called 'the director', who is on par with the experimenter in the sense that he knows what the experiment is about and follows the instructions. The director must give directions to the other person, who is the naïve or innocent participant (he does not know the purpose of the experiment). It is precisely the behaviour of the latter that we are interested in studying. The two subjects are separated by a 4 x 4 grid in which there are several objects (see Figure 3.5). Some of the objects that the participant sees, however, are not visible to the director. This information is known by both the director and participant. Therefore, from the participant's perspective, there are objects that he sees and he knows that the director cannot see. We call the stimuli that only the participant can see *distractors* and you will soon find out why. Imagine that the director asks the participant: 'Please, give me the small car.' From the participant's perspective, he can see three cars, one small, one medium, and one large and, therefore, he should give the director the smallest one. However, here is the trick: from the perspective of the director, the small car is covered and, therefore, he cannot see it.

The naïve participant knows that from his perspective, the director cannot see the small car, he can only see the big one and the medium one. Thus, when the director asks for the small car, it is impossible for him to refer to the smallest of the three, since he can only see two, the big one and the medium one. Therefore, the director must be referring to the medium car, which, from his point of view, is the small one. Basically, the idea is that the director sees fewer things than the participant, and he knows this. The question then is: what will the participant do when the director asks him to give him the small car? If he were able to take the director's perspective, he would have to give him the medium car. Something to the

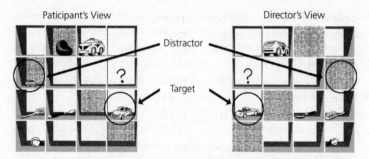

Paticipant's View · Director's View · Distractor · Target

Figure 3.5. Position of objects from the perspective of the researcher and the naïve participant. As can be seen, there are some objects that are only visible to the participant and therefore it is impossible for the director to take them into account. That is, he can never refer to them, because he does not know that they exist.

effect of the following must be going through his mind: 'The director is asking me to give him the small car. I see that there are three cars and, therefore, I should give him the smallest of the three. Of course, I also know that the director only sees two cars, the big one and the medium one, and therefore, when the director asks me for the small one, he is referring to the one I see as medium.' Easy, right? But if the participant suffers an egocentric bias and does not adopt the perspective of the other, he will give the director the smallest car of the three, because from his point of view (and that is the crucial thing, from *his* point of view) this is what the director is asking for.

Children who are naïve participants have problems performing this task. Very often they show egocentric bias and give the object in question from their perspective and not that of their interlocutor. And here comes the interesting discovery: it turns out that monolingual children between four and six years old choose the wrong object about 50 per cent of the time, whereas those children who have grown up in a bilingual environment do so about 20 per cent of the time. Furthermore, regardless of whether the children performed the task adequately or not (whether they gave the right object to the director from the latter's point of view), the authors evaluated where their gaze was directed right after hearing the instruction. That is, they measured their first reaction. As it turns out, monolingual children tended to look more often at distracting objects. That is, their first evaluation of the situation was egocentric. But there is another surprise yet: better performance by the bilingual children occurred regardless of whether they currently used the two languages or not. It was enough to have grown up in a bilingual context to show this better performance.

These results suggest that children who grow up in contact with two languages develop the ability to put themselves in each other's shoes earlier and change their perspective to that of their partner. So, the next time you ask for directions, let's hope you ask a bilingual person.

The origin of this greater capacity to put oneself in the perspective of the interlocutor may be related to the earlier development of the ability to see the intentions of the other, or what we sometimes call *mind-reading*. Do not panic, this has nothing to do with clairvoyance, fortune-tellers on TV, or other hoaxes like that. We are all reading the minds of others continuously. We know that others have intentions, desires, and knowledge, and that they are private and perhaps different from our own. It is essential to develop, for example, empathy, or the ability to put yourself in the shoes of others. To put it simply: we know that others have minds like ours and that the information in them can be shared or not. The development during childhood of what has been more formally called 'theory of mind' is fundamental. It not only allows the ability to empathize and is crucial for socialization, but also permits, among other things, the ability to lie. As one of my teachers said: rejoice when your son lies to you, but only the first time.

It turns out that there is evidence to suggest that children exposed to two languages show theory of mind development earlier than monolinguals. But how is it possible to explore children's ability to read the minds of others? Let's take a look. In a study conducted in Italy, the false-belief test was used. In this experiment, the researcher explains the following story to the children: 'A boy puts a chocolate bar in a red container in the kitchen and then returns to his room to play. While the boy is playing, the mother enters the kitchen and moves the chocolate bar into a cardboard box.' Then the researcher asks: 'When the boy returns to the kitchen for the chocolate bar, where do you think he will look for it?' The answer for you is clear: in the place where he left it, the red container. In order to answer correctly, the participant must understand that for the child in the story, everything that has happened in the kitchen when he was playing in his room is unknown and, therefore, he would have to look for the chocolate bar in the place where he left it, in the red container, and not where it is at the moment, in the cardboard box. But to respond to this answer, the participant has to put himself in the place of the child in the story, he has to contemplate the perspective of the other. He has to differentiate what he knows from what the child in the story can know. In short, the participant has to be able to separate himself from the content of his own mind in order to read the mind of the other. It turns out that many children up to the age of four fail in this task and believe that the child will look for the chocolate bar where it is actually located at that moment, in the cardboard box. It's as if they thought: 'I know the chocolate bar is in the box and, therefore, the child in the story will also look for it there.' The results of this study showed that at approximately four years old, around 60 per cent of Romanian-Hungarian bilingual children performed the task correctly, whereas only 25 per cent of Romanian monolingual children responded successfully. Surprising, isn't it? Bilingual children seem to develop a 'theory of mind' earlier than monolinguals.

But where does the effect of bilingualism come from in the development of the capacity to put oneself in the place of the other? Perhaps it is due to the need for the bilingual baby to differentiate between the sounds made by the parents. In other words, if the child has seen his parents speak in different languages from a very young age, maybe that will have helped him to hypothesize that his parents' minds are different to a certain extent. And if his parents' minds are different, then his must also be. It is precisely this that could help the development of theory of mind. But this is only a hypothesis.

Fortunately, as adults we are all able to pass the false-belief test. This does not mean, however, that we all have the same ability to take perspectives different from our own and put ourselves in the place of others. I am sure that I do not have to present any experimental data to convince you that there are more and less empathic people in the world. But I think you will be surprised to learn that in more complex tasks about false beliefs, bilingualism still seems to have an effect in adulthood by reducing the egocentric bias.

BILINGUAL VERSUS
MONOLINGUAL BRAINS

This brings us to the evidence we have about how the bilingual experience can sculpt the anatomy and the functioning of some cerebral structures and circuits.

Any learning that we carry out has an effect on our brain. Learning is possible thanks to the plasticity of the brain, which involves the creation of new connections between neurons as a consequence of storing new information. Throughout life we learn factual or declarative information about the world around us: words, phone numbers, land masses, the ingredients of an omelette, our city's streets, the rankings of our favourite teams, the names of the elements of the periodic table, that cod and rice is better if the rice has peas in it, and so on. This type of information is what we often say is *learned by heart*, and we see how, as some neurodegenerative diseases progress, it disappears. But we also learn how to do things: to walk, cycle, swim, drive a car, speak and read, and so on. This is what we call procedural information, which is what allows us to carry out highly automated activities.

Learning a language involves the absorption of these two different types of information, since on the one hand, we have to acquire the lexical items (vocabulary) and on the other, the grammatical processes to combine them (syntax). But how does the acquisition and use of two languages affect the brain? In other words, is there a difference between the brain of bilinguals and that of monolinguals in terms of the neural network responsible for processing language(s)?

Neuroimaging techniques have been fundamental in answering this question. At the functional level, several studies have shown that there are differences between the activation levels of certain areas of the brain when bilingual and monolingual individuals process their first language. It is important that it is the dominant one, because what interests us here is not so much the processing difference between a first and second language (we already discussed that in Chapter 2), but to what extent the processing of the first language is different among bilinguals and monolinguals. Returning to the analogy of the sports practised by David, squash and tennis, the question is how the learning of two sports affects the cortical representation of the one that was first known, that is, how learning squash affects the cortical representation of tennis.

Perhaps the most complete study on this issue was conducted by Cathy Price and her collaborators at University College London, in which the brain activities of highly proficient Greek-English bilinguals and monolingual English speakers were studied across various linguistic tasks. The results showed that brain activity in language comprehension tasks, such as speech perception, was very similar for both groups. However, those tasks that involved the language production system, such as picture naming or reading aloud, *did* reveal differences. Specifically, bilinguals showed greater activation in five areas of the brain located in the left frontal and temporal lobes. I don't want to bore you with the details about the specific interpretation the authors make about each area. I will only mention that other studies suggest that these same areas of the brain are related to effects of frequency of use and linguistic control. What is important to note is that, at least in this study, no significant differences were observed in the areas that were activated in bilinguals and monolinguals. To a large extent they were the same, although, yes, with greater intensity for bilinguals. These results were interpreted by the authors as evidence that, either due to the lower use of each of the languages or the need to control interferences (or both reasons), bilingual speakers require a certain overexertion during speech production compared to monolinguals. Other studies carried out with different groups have shown similar patterns and, in fact, are even stronger when second-language proficiency is low. These results suggest that the learning and use of a second language does not radically affect the brain representation of the first language, but it does affect the effort required for its processing, especially when speaking.

However, other studies have shown the existence of certain unique characteristics related to bilingualism. For example, in a study conducted at Jaume I University by César Ávila and colleagues, the brain activity of Spanish-Catalan bilinguals was compared to that of Spanish monolinguals while performing various tasks in Spanish, their first language. Similar to what we saw above, the differences between the groups were very small when the activity involved auditory word comprehension. Yet when participants were asked to name drawings, it was observed that bilinguals tended to use a wider brain network than monolinguals. In other words, bilinguals incorporated areas of the brain that were not deeply related to linguistic processing. This could support the existence of certain areas of the brain, located mainly in prefrontal areas, that only bilinguals use during speech.

These results reveal that the cortical representation of a bilingual's first language is, in general, quite similar to that of the monolingual one. The classic areas where language processing takes place are involved in both cases. But this does not mean that bilingualism does not affect how those areas are utilized and, as we have seen in this case, it is possible that some of these areas simply have to 'work harder'. So it seems premature to discard the idea that there may be certain areas that are activated more in bilinguals. And it is very possible that these areas have to do with control processes and not so much with the representation of linguistic knowledge.

STRUCTURAL CHANGES

In the previous section we described studies that measure brain activity during different linguistic tasks. Learning and using two languages seems to have not only functional consequences, but also implications for brain structure. By 'brain structure' I am referring to and encompassing the density or volume of basically two kinds of tissue: *grey matter* and *white matter*. Simply defined, the density of grey matter is the number of neuronal bodies and synapses present in a given space of the cerebral cortex. White matter refers to nerve fibres covered with myelin, basically those that include myelinated axons. These fibres are fundamental for transmitting information between neurons, and myelin acts as an insulator that allows nerve impulses to be efficiently transmitted (like the plastic that covers an electric cable). To put it another way (and neurologists, please don't get upset by the analogy): grey matter is that which computes information and white matter is the cable that is responsible for transmitting that information from one place to another.

It turns out that the density of grey and white matter can be altered by learning a new skill. For example, a study published in *Nature* showed that training in juggling resulted in various changes in grey matter of areas of the brain related to the processing and storage of complex visual-motor information. Other studies have shown that such modification occurs after only one week of training. Another more recent study published in *Nature Neuroscience* identified the effects of juggling training on the architecture of white matter. Learning modifies the brain, so in a way, we could say that knowledge does occupy a place, or at least modifies the structure of that place in terms of brain architecture.

In fact, we also know that it is not necessary to participate in training to alter the structure of our brain, and that daily activities can also result in some modifications. Perhaps the most well-known case of this issue is a study in which the brain structures of a group of taxi drivers in London with an average of fourteen years of experience were compared with those of a control group that, although it shared many other variables, did not have experience of taxi driving. Keep in mind that at the time the study was conducted, the use of technology to help with directions was not as widespread as it is today and, therefore, the drivers needed to learn the London map by heart. The authors observed that, curiously, taxi drivers had a greater volume of grey matter in the area closely related to the storage of spatial representations, the anterior part of the left and right hippocampus. In addition, this greater volume correlated with the years of experience at the wheel: more years, more volume. That is, with more experience, there was more grey matter. These results suggest that activities that we carry out daily have an effect on brain structure. Our behaviour and learning sculpt the brain.

The question is whether the acquisition of two languages affects in some way the cerebral anatomy or, if you will, the brain's *structural architecture*. Notice that I use the term structural architecture to differentiate it from the brain's *functional architecture*, which we discussed in the previous section. In the first study that analysed this question, Andrea Mechelli and her colleagues compared the structure of certain areas of the brains of monolingual and bilingual speakers, and showed that one area in particular, the left inferior parietal lobe, had a greater density of grey matter in bilinguals than in monolinguals. This happened both when the second language had been learned in childhood and when it was learned later on. In addition, bilingual individuals with a more extensive vocabulary in the second language also showed greater density in that area of the brain. These results led the authors to suggest that learning second-language vocabulary has consequences on the development of the grey matter of that particular area of the brain.

The plasticity of certain areas is not only reflected in the acquisition of new words, but also sounds, as suggested by the observation that multilingual speakers have a greater density of grey matter in the area involved in articulation and phonological processes, namely the left putamen. Thus, a more extensive phonological repertoire and the need to control the articulatory movements of each language would affect the structure of the areas responsible for these representations.

Studies that compare the brain structure of monolinguals and bilinguals face a problem when they try to give a causal interpretation to the results. It's like the chicken or the egg problem. We cannot be sure whether the bilingual experience sculpts the brain in a certain way or whether those individuals with a special type of architecture are the most prepared to learn a language and, therefore, have more facility to be bilingual. If this were so, growing up in a bilingual environment would not affect brain structure, there would simply be a relationship between both variables, but not causal. To explain it in more practical terms: if we compared the average height of basketball players to that of soccer players, we would see that it is different, but that does not mean that practising basketball makes people taller or that playing soccer makes them shorter. It is precisely because they are tall that they play basketball and, continuing with the analogy, given that some individuals show larger grey-matter density in the relevant areas, they are more able to learn a second language successfully and become bilingual.

There are two ways to solve the causal interpretation problem. The first is to evaluate individuals who are bilingual not because they have learned the second language in a regulated manner (e.g. in school) but because they were born or lived in bilingual environments. That is to say, a child born in a family where English and Spanish are spoken will go beyond their brain architecture and learn the two languages;

he will know how to play basketball (if that is what his parents play), regardless of his height. Therefore, if we find differences in the brain structure of this type of bilingual compared to that of monolinguals, we cannot attribute them to bilingualism through regulated acquisition, but to the result of the bilingual experience. Let's discuss a couple of studies.

In a study conducted with Spanish-Catalan bilinguals, whose bilingualism was due simply to the environment where they had grown up, it was observed that the volume of the left Heschl's gyrus was greater than in monolingual speakers, both for grey and white matter. This area of the brain is related to phonological processing and therefore the authors concluded that experience with two languages of relatively different sounds affects the development of the areas responsible for their processing. But this region is not the only one whose volume is increased. In a study focusing on a similar population of Spanish-Catalan bilinguals, it was observed that differences in grey matter occur even in deep areas of the brain, areas that, until not long ago, were thought to have less participation in such complex processes as language production or comprehension. Today we know that these areas, which include the basal ganglia and the thalamus, are involved in the articulation of speech sounds, among other things (see Plate 2). Bilingual individuals make use of these areas to a greater extent since they must learn to produce a greater number of different sounds.

The other way to determine the causal relationship between the bilingual experience and changes in the brain is to conduct studies in which the effect of language learning on brain structure is measured. These studies tend to present their own challenges, since they are ideally longitudinal and, therefore, require analyses of participants at different time points. In one of these studies, the changes experienced by native speakers of English were evaluated during a second-language German immersion experience. The brain measurements related to learning the second language were taken at the beginning of their stay in a German-speaking environment and once again five months later. A correlation was observed between how much they had learned from the starting point and the change in grey matter density in an area of the brain related to language, the left inferior frontal gyrus.

Individuals who learned more German showed a greater change in the density of grey matter in this area. Note that this relationship is independent of the final level of competence acquired in the second language; it points to the difference between the level at which learning was started and the level at which it ended, which suggests that the important thing is *how much* the participants had improved and not *up to what point* they improved. There you have it, then: if you send your child abroad, expect changes not only in their meal times but also in their brain's grey matter.

Other studies have also analysed how the age of acquisition of a second language can affect brain structure. In one of these works, a curious and interesting pattern was observed. Those bilinguals who had learned a second language after childhood showed, compared to monolinguals, more grey matter in the left frontal gyrus and less in the corresponding area of the right hemisphere. In addition, surprisingly, this effect was not found in individuals who learned two languages from birth, who did not show differences from monolinguals.

The bilingual experience also seems to affect the development of white matter, but the results of the various studies regarding this claim are a little less conclusive. Thus, while some experiments show the existence of changes in the corpus callosum (the fibres that connect the two hemispheres), others have found differences in the occipitofrontal fasciculus. There are yet other studies, which we will discuss in the next chapter, that have observed such difference in other brain fibres.

Finally, it is important to note, as some researchers such as Manuel Carreiras at the Basque Centre on Cognition, Brain, and Language have recently pointed out, that the evidence we currently have about how bilingualism sculpts the brain is somewhat inconclusive and confusing. In addition to the fact that the results of the different studies are inconsistent within and among themselves, there are not too many published works that provide a more reliable and accurate view of the areas of the brain affected by bilingualism. This is a problem, it's true, but also an opportunity to continue exploring the interaction between a daily activity such as speaking in two languages and brain plasticity. I have no doubt that in the next few years we will make more progress.

4

Mental Gymnastics

I am writing these lines in a hotel in Manhattan as I am preparing to give a talk about how bilingualism can affect the development of the attentional system. If you have ever walked around New York, or any other big city, you will have experienced the amount of stimuli that are continually calling for your attention: lights of the advertising panels, traffic noises, firetruck sirens, all the different people with whom you come into contact and, of course, the smells wafting from food carts. All of them are stimuli that insistently attract your attention. The city is quite an experience for all of our senses, but also a challenge for the attentional system, which has to ensure that, despite all that distraction, I arrive on time for my talk. It's fun, but also tiring. I'm telling you all this because this whole chapter is about how bilingualism can be a factor that affects our attention, among other cognitive abilities.

The theme of the conference I was attending has become increasingly popular both in the scientific community and in the media. It focused on the extent to which the continuous use of two languages has consequences for the executive control system; and, more specifically, up to what point these consequences can produce benefits with regard to attention. If such benefits are real, we can say that bilingualism has consequences not only at a social, cultural, and economic level, but also in the development of crucial functions such as executive control. As you can imagine, such evidence would be diametrically opposed to the idea that being bilingual can lead to cognitive problems, a theory that was popular fifty years ago.

This hypothesis is based on the idea that the linguistic control that bilinguals must exercise involves processes common to the executive control system. So when bilinguals process language, they are also exercising the processes and the corresponding brain structures that form part of central executive functions. Let's look at an example. In Chapter 2 we saw that, according to certain models, when bilinguals use one of their languages, there is certain activation of the representations of the other language, which we have called the *language not in use* (remember the experiment with Chinese-English bilinguals). We have also explained that in order to avoid interference from the language not in use, it seems that inhibitory mechanisms are put into play that make their corresponding representations not act as potential competitors. That's how we manage not to mutter random things all the time and mix languages without wanting to. The hypothesis predicts that these inhibitory mechanisms are the same ones that are put into play when I go to my talk and do not get distracted by all the stimuli that catch my attention on the way. That is, the smell of hot dogs and the sound of firetruck sirens – attractive stimuli but irrelevant to my goal of getting to the talk – are ignored or inhibited by my executive control system. And those inhibitory processes are the same ones that bilinguals use when controlling their languages. If this hypothesis were true and since humans are just talking heads who spend so much of their time using language, then all the exercising (and the occasional juggling) that bilinguals do to control their languages could imply a more efficient attentional system. Nice hypothesis, don't you think?

Before going on to describe how this hypothesis is being evaluated experimentally, it is important to clarify a few things. First, it is one thing for bilingualism to have effects on the brain networks that are involved in the executive control system, but it is quite another thing to find that these effects result in a tangible advantage for its operation. In my view, when we talk about advantages, they should be measurable behaviourally, that is, if the advantages result in better performance on tasks that involve the executive control system. If a bilingual is able to walk around Manhattan and get distracted less than a monolingual, and can get to the talk without having to rush, that would be a tangible advantage. If, on the contrary, they both come to the talk with equal punctuality but using relatively different brain networks, then I'm not sure we should consider it an attentional advantage. Of course, this last case would continue to be an interesting and relevant fact from a *theoretical* point of view, but we could not claim that this is a *behavioural* advantage.

Second, it is important to determine the magnitude of the effect of bilingualism. Most of the activities that we usually do involve the executive control system, from driving a car to making coffee and talking on the phone at the same time. We are constantly exercising this system. So it is important to determine how and to what extent bilingualism can imply an additional exercise that results in more efficient executive functioning. We will return to these two points later.

AVOIDING INTERFERENCE

According to surveys, 80 per cent of people think that they drive better than average. But that can't be true. Unless there has been a huge sampling error, which is not the case, or we only asked the best drivers, the number reveals just how confident we are in ourselves. (By the way, the same result is obtained when people are asked about their lovemaking skills.) Driving is a challenge for the attentional system: we have to keep in mind the place where we want to go, ignore irrelevant information that can confuse us, react quickly when there is danger on the road, and so on. If you have any doubts, just take a look at Figure 4.1 and think about what you would do in that situation. When the act of driving becomes more automatic, it gives us the impression that we are not paying attention to anything mentioned above, but we do, unconsciously. The same happens when bilingual speakers have a conversation. As we have seen in Chapters 2 and 3, a series of control mechanisms that allow fluent speech in the desired language are put into play, avoiding interference from the other language. Let's look at several studies evaluating the effect this linguistic control can have not on driving, but on the attentional processes that are related to it.

From the previous chapters, I'm sure you noticed that if cognitive

Figure 4.1. A real example of how we find contradictory information continuously. This contradiction has to be resolved by the supervisory attentional system.

psychologists are good at something, it is at creating ingenious experimental situations to address complex issues. Let's look at the next experimental paradigm that results in the so-called *Simon effect*, which bears the name of the scientist who discovered it in the 1960s. The experiment is simple: red or green circles are shown on a computer screen, one after another, and the participant is asked to press a key with the right hand (the *m* key, for example) when a green circle appears, and to press a key with the left hand (the *z* key, for example) when a red one appears. That's all. Kind of boring and easy. How is this ingenious? The trick is in the following: circles can appear in the left, centre, or right portion of the screen. In principle, this dimension, that is, where the circle appears, is irrelevant to the participant's task, in which he simply has to decide with which hand to respond depending on the colour of the circle. However, when he has to press the key with the right hand (green circle) but the circle appears on the left side of the screen, response times are slower than when that same stimulus appears on the right side (and the same thing happens when the red circle appears on the right compared to when it does on the left). It's as if the participant could not ignore the side on which the circle appears and, when that side does not match the hand with which he must press the key, there is a conflict that he must resolve which requires more time. That's the Simon effect: the difference in the time it takes to give the answer when the stimulus appears incongruent to the response hand and when it appears congruent. Has anyone ever told you that you have to turn right but accidently showed you with his left hand? This is the Simon effect in real life.

Studies by Ellen Bialystok at York University in Toronto revealed that bilingual speakers showed a reduced Simon effect compared to monolinguals. That is, the effect of the conflict generated by the incongruent conditions was smaller for bilinguals. This difference was found among participants who were thirty years old and the effect was ever larger after sixty years old. Age certainly had some-thing to do with it given that after sixty years the effect increased, but this applied much more to the monolingual participants. These results suggest that the bilingual experience affects the ability with which participants focus their attention or are able to resolve the conflict between relevant and irrelevant information. Crucially, this is found in a spatial task that has almost nothing to do with the lin-guistic system, which reveals the effect of bilingualism on the general executive control system (the one we use when driving). By the way, this is another one of the experiments you can do with your friends at home. Ask a friend to raise his right hand when you show two fin-gers, and the left hand when you show one finger. Then start: show one finger with the left hand, then two with the same hand, two with the right hand, and so on, mixing congruent and incongruent stimuli (make sure that your arms are wide open when showing fingers). You will soon see that your friend is confused by incongruent stimuli, especially if you've had a few glasses of wine.

Many other studies have shown similar results with respect to the improvement of conflict resolution in bilingual individuals, although there are also certain doubts, as we will see later, on the replicability of these results. Let's see another example, which we will return to in later sections: a study we published in *Cognition* that was carried out in 2008 in my former laboratory at the University of Barcelona in which we set out to see whether the bilingual experience has effects on conflict resolution when individuals are at their *peak point* of attentional capacity. What do we mean by peak point? One of the areas of the brain that matures more slowly is the prefrontal cortex, which is very much directly involved with attentional control. This area develops up until puberty and reaches its optimal functioning in the twenties . . . and then, bad news, it begins to slowly decline in the thirties. Maybe that's why, among other things, the best years for athletes are usually from twenty-five to thirty, when they are able to overcome conflicts with the greatest speed. In our study, we sought to see whether bilingualism had some positive effect on the attentional capacity of twenty- to thirty-year-olds. For this, we recruited 200 participants – 100 Catalan-Spanish bilinguals and 100 Spanish monolinguals from different Spanish universities – and asked them to perform the *flanker task*. In this experiment, the participant sees a stimulus of the type →→ → →→, and is asked to say in what direction the middle arrow (which we will call *target stimulus*) is pointing, while ignoring the presence of the arrows that are on both sides (the *distractor flanks*). The trick is that there are congruent stimuli, like the one that we just saw, in which the flanks point in the same direction as the target stimulus, and there are incongruent stimuli, where the flanks point in the other direction (←← → ←←).

As in the Simon effect, presented above, the answers are faster and more precise for congruent stimuli than for incongruent stimuli.* As you can see in Figure 4.2, the results of the study showed that the bilingual individuals suffered less interference than monolinguals in all the blocks of the experiment, although these effects were greater in the first two blocks. This was the first published result that showed a positive effect of bilingualism on conflict resolution among young adults.

Factors such as age of acquisition of the second language, acquired competence, the daily use of both languages, and even the frequency with which bilingual conversations are held, have also been explored to see whether these factors affect the presence and magnitude of the positive effect of bilingualism on attentional control.

* You can try the experiment for yourself by visiting http://cognitivefun.net/test/6.

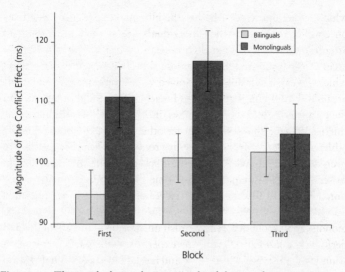

Figure 4.2. The graph shows the magnitude of the interference among bilingual and monolingual speakers in the flanker task. The bigger the bars, the greater the conflict experienced.

Given the information that we have at present, it seems that the most important factor is the regular use of the two languages. That is, the positive effect of bilingualism on conflict resolution would be linked not so much to the degree of competence with which the second language is spoken, but to how often it is used. This regular use would imply the activation of linguistic control processes among bilinguals, who in turn would exercise executive control processes. So if you want to enjoy this effect, do not worry too much about how well you know a second language, but do practise it frequently.

These are just two of the many studies that have compared the skills of bilinguals and monolinguals when it comes to resolving conflicts or interference in tasks that do not involve language, or involve it very little. However, not everything is so simple and, as we will see, in recent years several researchers have questioned both the replicability of the results of these experiments and the existence of advantages associated with bilingualism.

MULTITASKING

We live in the era of multitasking. We write a text while talking on the phone with a friend, we go over the bills that are going to have to be paid this month while we make coffee, and we chat while we have dinner. Many of these activities are done in parallel, that is, at the same time, which requires that our focus changes from one to the other. We call this *task switching*, which is difficult and costly, and can sometimes cause us to make mistakes. This skill can also be trained, although it reaches its peak in our twenties. Maybe that's why I'm still beating my son on video games from time to time; he has turned sixteen and I have to hurry: time is running out.

This attentional capacity has often been related to the ability to switch languages in bilingual individuals. Once again, it has to do with the fact that bilinguals have to change or control the language according to the interlocutor, which activates circuits similar to those involved in task switching in the general sense and, therefore, grants bilinguals a certain advantage over monolinguals. Remember the family in Chapter 2, for example, whose members changed languages continuously, although not randomly. So we could say that bilinguals are better at multitasking.

This hypothesis has been put to the test in several studies that have carried out activities similar to the one described in Chapter 2. As you'll recall, the participants had to say the name of what was represented in various drawings in their two languages according to the colour of the border in which they appeared. If it was red, they named it in language A, and if it was blue, in language B. The difference in the speed and accuracy of the responses between switch trials and repetition trials is what we have called *language-switching cost*. So, you can use this same design, or something similar, to measure the ability to switch tasks in situations that do not involve language.

One of the first studies to address this issue adapted this activity for children. The participants were shown a series of coloured round and square cards, blue and red, and were asked to classify them according to their colour (blue cards on the right, red cards on the left). Once that task was finished, the cards were shuffled, and they were asked to classify the cards according to their shape regardless of the colour (circles on the right, squares on the left). That is, they were asked to switch tasks or, if you will, to switch classification criteria (first colour, then shape). A rather simple experiment, you might think, but remember that the participants were children between five and six years old. Those who spoke two languages, in this case Cantonese and English, performed the task with better results than monolingual English children. Better results, in this case, means that while there were no differences between the groups in the first task, which shows that they understood it perfectly, there were differences in the second task. So when the children had to change the classification criteria, that is, the task, the monolinguals were mistaken more often than the bilinguals.

Subsequent studies conducted by Anat Prior at the University of Haifa and Tamar Gollan at the University of California, San Diego, showed that this advantage in bilingual children is also observed in young adults, which suggests that effects of bilingualism in this cognitive domain are present throughout various phases of development. In addition, in these studies, the bilingual advantage was related to the frequency with which they code-switch between their languages.

In this area of research, the more surprising result is the one that Agnes Kovacs and Jacques Mehler observed when they explored the ability to switch tasks among monolingual babies and seven-month-old bilinguals . . . yes, you can study how babies change tasks! To do so, the researchers showed the babies two blank boxes on the computer screen and, between them, a fixation drawing that in very little time was replaced by a triangle. The triangle remained for one second while nothing else happened. Then, another drawing appeared to attract the baby's attention, always in the box on the left. This was repeated nine times (fixation drawing – triangle – one second where nothing happens – attractor drawing, always in left box). What was interesting here was observing where the child would look during the one second that passes between the appearance of the triangle and that of the attractor drawing. If the baby were able to realize that the triangle always preceded the appearance of the attractor drawing, then perhaps he or she would show the tendency to anticipate the location of the latter. Indeed, after several trials the babies were able to anticipate the location of the stimulus, and we know it because during the one second, they oriented their gaze towards the left box where the attractor picture was going to appear. In other words, they more often directed their attention towards the left before the drawing appeared, as if to say: 'I direct my eyes because that is where that drawing that I like so much will pop up in a little bit.' And, here's the trick. After those nine trials, instead of a triangle, there was a circle, and after it, another attractor drawing but in the opposite box, that is, the one on the right. These trials, which were shown nine times, are called *post-switch* trials, since the location of the stimulus has switched sides. What was recorded here was the same as before, that is, where the child looked during the time that elapsed between the disappearance of the clue (in this case, the circle) and the appearance of the attractor stimulus, that is, when only the blank boxes were shown on the screen.

Would the babies be capable of reorienting their attention and anticipate the new position of the attractor drawing? In the first trials, the majority of the babies were not able to; in fact, and to be fair, in the first trial, they still could not know what change was going to happen. However, little by little, the babies began reorienting their attention and anticipating where the stimulus would appear. It is as if they thought: 'Okay, I got it, now the neat object will appear on the right.' *Voilà*, they were able to change tasks, or at least change criteria! Of course, at seven months old, the babies who *were* able to refocus their gaze were the ones growing up in a bilingual environment (in this case, most of them were Slovenian-Italian bilinguals). The Italian monolingual babies persisted on continuing to watch the box on the left, as if they were stuck on the first criterion and were not yet flexible enough to change it.

This study is important because it shows the effect of bilingualism on the cognitive flexibility of very young babies. Bilingualism helps develop an attentional system sufficiently flexible to: 1) allow it to inhibit the answer learned in the first trials, and 2) help it to update predictions according to the new task demands. The results matter because these babies still do not speak, so the positive effects of bilingualism for the attentional system do not necessarily, or exclusively, come from the linguistic control exercised during speech production.

There must be something more. Maybe the mental exercise that the babies are doing when trying to differentiate the sounds that come out of their parents' mouths is what is giving them this mental flexibility.

NOTHING IS EVER EASY

Several researchers have recently expressed doubts about the reliability of studies that show bilingualism's positive effects on the development of the executive control system. Let's review some of their claims because I think it's a good scientific exercise that's not only applicable to this topic but to science in general, and to the social sciences in particular.

Some have wondered whether science journals are more interested in studies showing differences between bilinguals and monolinguals. This bias means that not all studies have the same probability of being published in scientific journals due to editorial trends and, therefore, remain unknown to researchers and society in general. At best, these studies are presented at conferences but are not published. Of course, this is not necessarily bad, because not all studies are equally rigorous methodologically. For example, a study comparing the performance of bilinguals and monolinguals on attention tasks in which the samples were not balanced in, for example, the age of the participants, would have little chance of being published, and for good reason. Therefore, it is logical and desirable that editors and reviewers of scientific articles tend to positively evaluate studies that offer more solid proof rather than those that do not, and that it's the first ones which are usually published. That is the work of reviewers and editors, a cumbersome job, which takes a lot of time, is not well paid, sometimes creates enemies, and so on. I'm speaking from experience, since I have played, and continue to play, all these roles: those of an author, a reviewer, and an editor. While peer review isn't perfect, it is the best way scientists have found so far to decide which studies should be published.

In fact, one of the problems in peer review is the existence of publication biases that are not due to the experimental quality of the studies, but rather to the results they obtain. Peer review tends to value more positively those studies that, independent of their experimental quality, show significant differences between the experimental conditions, that is, the work is judged by the observed result and not only by its scientific rigour. For example, suppose a laboratory team has developed a medicine and they want to test its effectiveness, and suppose also that they perform the experiment correctly. The result can be positive or negative, that is, the medicine either works or not. Well, the probability that their study will be published in a prestigious journal increases based on that result, being higher if this is positive. If the experiment does not show differences between the group that was administered the medicine and the control group that didn't take the medicine, we could say: 'Well, this has not worked, that's too bad, maybe we have not controlled something or we have made an error in some of the measures, or who knows . . .' What appears to be the case is that we believe that nothing has been learned. On the contrary, if the result is positive and the experiment is done well, the possibilities increase not only of the outcome being published but also of it being featured in the newspapers. But have we really not learned anything from that negative result? I think we have learned something, and although you have to proceed with caution, I do not see why the probabilities of publishing those studies should be smaller. Publishing that negative result, at the very least, could save other researchers time and money so they do not conduct a study testing that same hypothesis in the future. If they know that the medicine was already tested and did not work, maybe they can dedicate their resources to something else. That's what would happen in an ideal world.*

I have no doubt that when it comes to attentional advantages associated with bilingualism, there *is* a publication bias, and to say that the same thing happens in other fields does not comfort me

---

* Unfortunately, we do not have a direct way of knowing whether there is a publication bias and if so, what its magnitude is, but there are indirect ways of detecting it by using statistics. If you are interested in these issues and how the world of general science works, I recommend Ben Goldacre's *Bad Science*. Sometimes the problem is not the bias caused by the editor or reviewers, but by the authors themselves, who file away their results in the 'failed experiments drawer'.

much. But as the saying goes, 'two in distress makes sorrow less'. That said, and to avoid confusion, the fact that there is a publication bias does not necessarily negate the existence of the effect of bilingualism on the attentional system. There is hope, as there are ways to reduce that bias. What we need to do is to input the methodological properties of the experiment into a database, including the design and the population that it studies. A condition for its subsequent publication is to have recorded it before it is carried out, so we as researchers are very careful. This policy is already followed in many clinical studies, although unfortunately not in all. If we register the studies, we can know how many of them are eventually published and how many are not, and as such, we can make an estimate of those that have been submitted and accepted, and those that have been filed away in the 'failed experiments' drawer. We could always contact the researchers and ask them to share their observations, whatever they may have been.

Let's go back to bilingualism. Some researchers have pressed hard with their criticisms and have questioned the reliability and reproducibility of the results that show a positive effect of bilingualism on the attentional system. They have been trying to replicate the results of studies already published. Here are two examples.

The first is an experiment led by my colleagues at the Basque Centre on Cognition, Brain, and Language, evaluating the performance of 500 children aged eight to eleven who were either Basque-Spanish bilinguals or monolinguals of Spanish. The researchers were careful to apply balance to the two groups across several variables that could have affected performance on the activities carried out, which were two Stroop-type tasks, in which one must resolve conflicts generated by irrelevant information, similar to the conflicts found in the Simon effect; and the flanker task, which we discussed above. In none of these tasks did the researchers observe differences between the performance of bilingual and monolingual children. This result led the authors to write an article whose title says it all: 'The inhibitory advantage in bilingual children revisited: Myth or reality?'

The second study was conducted by Mireia Hernández in my laboratory at the Pompeu Fabra University. In this study, we wanted to understand better the mechanisms responsible for the advantage associated with bilingualism during task switching, which we discussed in the previous section. For this, we used several experimental procedures, all of which involved task switching. We even replicated exactly the same experimental design used by other authors who had found higher task-switching costs among monolinguals. Although we were able to detect certain effects among the bilingual individuals, we failed in our attempt to replicate the reduction of task-switching costs associated with bilingualism. And it was not because we weren't making an effort, since we evaluated the performance of almost 145 Spanish-Catalan bilinguals and 145 Spanish monolinguals. When we compared the results, the distribution of the magnitude of task switching was virtually identical for both groups.

There are also more theoretical questions about our continuous use of the attentional system, regardless of whether we are bilingual or not. Irrespective of whether we speak two languages, we are using our executive control system continuously, so bilingualism has very little to add to its operation or development. It is as if we've reached the ceiling in our use of that system and, however much we exercise it, our performance does not improve significantly. In baking terms: I cannot make a cake any better, I make it so well that no matter how hard I try, it just doesn't get any tastier. According to these researchers, this ceiling effect would account for why the detection of the effects of bilingualism on the attentional system seem to be relatively unstable.

In my view, most of these criticisms have some validity. The question is, what do we do now? Which studies do we go with? With those that show an effect of bilingualism or with those that do not? Sometimes even in the same laboratory, positive effects are observed in some activities but not in others. This is an empirical issue that cannot be influenced by personal preferences or social interests. So the real question is not so much which studies we should go with, but rather how we develop a research programme that reliably allows us to identify the effects of bilingualism on the executive control system. And, in fact, researchers are trying to understand what variables and experimental contexts can favour the detection of bilingual effects. From my point of view, it would help if we stopped focusing on the advantages collaterally offered by bilingualism and tried to describe how being bilingual modifies certain cognitive processes and their corresponding brain circuits, without worrying about behavioural advantages.

SCULPTING THE BRAIN

In the previous sections we have reviewed the question of whether bilingualism positively affects the attentional development of an individual at the behavioural level. We have seen that the situation is complex, and we still need more studies that allow us to reliably ensure that the use of two languages indeed confers a certain advantage. This does not mean, however, that the bilingual experience has no effect on the brain structures involved, regardless of whether this is a behavioural advantage. The question, then, is to what extent bilingualism sculpts the brain, and more specifically, the areas involved in attentional control. That is, how does a daily activity, such as the use of two languages, affect the structure and function of some cerebral circuits? Do you remember the consequences of driving a taxi in London for the development of certain parts of the hippocampus? The underlying argument is the same here.

With this objective in mind, we participated in an investigation led by Jubin Abutalebi at the San Raffaele hospital in Milan, the same hospital where almost twenty years ago we carried out the study exploring whether age of acquisition affected brain representation of a bilingual's languages. In this case, we wanted to evaluate the brain overlap in the performance of a linguistic control task and in an attentional control task not involving language. The idea was that this overlap would give us information about which areas of the brain were involved in both tasks. To do so, we asked German-Italian bilinguals and Italian monolinguals to perform two different activities. The bilinguals came from the region of South Tyrol, in which, for historical reasons, German and Italian are co-official languages.

One of the tasks would tell us about linguistic control; it was (you guessed it) a language-switching task: if a drawing appears within a red border, name it in language A, and if it appears in a blue border, name it in language B. But how could we measure linguistic control for the monolinguals? They, in effect, could not switch between languages, for obvious reasons. So we asked them to switch between grammatical category: if a drawing appeared within a red border, they should name the noun it represents (*broom*) and if it appeared in a blue border, they should name the action that is carried out by the object (*sweep*).

The other activity did not involve language and was the flanker task explained earlier (congruent stimulus: →→ → →→; incongruent stimulus: ←← → ←←). So in the first experiment, we measured brain activity caused by language switching or grammatical category switching and, in the second, we analysed brain activity caused by the non-linguistic conflict. In other words, we compared brain activity between: 1) switch trials and non-switch trials, and 2) incongruent stimuli and congruent stimuli. We then evaluated which areas of the brain were overlapping in both of these situations, that is, which areas were involved in language switching and which were involved in congruence effects.

One of these overlapping areas that we found and that, in fact, we expected to find, was the anterior cingulate cortex, which previous studies had related to cognitive control and conflict resolution. We were on the right track. This area responded with greater intensity when cognitive control was increased in both activities. But the activation observed for the non-linguistic conflict (the activity with the arrows) in this area was less for bilinguals than for monolinguals. Even though bilinguals suffered less conflict than monolinguals at the behavioural level, the brain energy necessary to solve the task was less (see Plate 3).

My colleagues and I did not settle for just learning about this functional architecture; we went a step further and analysed the anatomy of this area, the anterior cingulate cortex. We observed that the density of grey matter was greater in brains of bilinguals than of monolinguals. This result would indicate that the continued use of two languages has effects on brain structures involved in domain-general executive control, that is, in an attentional system that is not uniquely linked to a cognitive domain, whether linguistic or non-linguistic.

Other studies have observed that, in task-switching paradigms that involve language either very little or not at all, bilingual speakers seemed to activate a more distributed brain network than mono-lingual speakers, including areas related to linguistic control, like the inferior frontal gyrus in the left hemisphere.

In addition, the effects of bilingualism on the attentional network are not limited to the functioning and structure of grey matter; they also affect the robustness of white matter (see Plate 4). As noted in Chapter 3, white matter is that which connects the neurons to one another, as well as the different areas of the brain. Remember, they are the cables that transmit information. With age, these cables det-eriorate and the integrity of the white matter (of myelin in particular) decreases. It's kind of like when the cord that connects your ear-phones to your cell phone has a poor connection, which results in crackling noises. Information no longer circulates as efficiently, and this affects the individual's cognitive functions. And, in fact, the operation of the attentional system is one of the most affected by the decrease in white-matter integrity.

In one of the most interesting studies on this topic, Gigi Luk at Harvard University and her colleagues compared the white-matter integrity of bilinguals and monolinguals with an average age of sev-enty. This type of analysis focuses on brain structure so it is not necessary for participants to carry out any specific activity while their brains are being scanned. They are in a resting state. The results were quite interesting. Bilinguals maintained greater white-matter integrity in the corpus callosum, which corresponds to the nerve-fibre groups that connect the two cerebral hemispheres. This does not mean that it was the same as in twenty-year-olds, but it was greater than that of aged-matched monolinguals. In addition, the authors also measured, through more complex analyses, the extent to which the two areas were functionally connected (the degree to which both areas 'spoke to each other'). The results were consistent with earlier studies in that they showed a more distributed connec-tivity in certain circuits in bilinguals than in monolinguals.

What does it all mean? The authors argued that the bilingual experience results in a strengthening of the connectivity of white matter, and that this is responsible for the higher performance of bilingual individuals in attentional tasks. It's an interesting hypothesis, but it's a shame that in this study no data were obtained on participants' performance in these types of activities, and we do not know whether a greater integrity of the white matter would have implied better performance.

These results, among several others, indicate how flexible the brain can become, and how the activity of learning and using two languages during life has tangible effects on its organization and development. It's interesting because these effects are not limited to those parts of the brain that have traditionally been related to language processing, but they seem to extend also to other areas that have to do with attentional control.

Although we still have more to learn in terms of the consequences that these changes related to bilingualism have on cognitive performance, these observations open an interesting hypothesis: could the bilingual experience affect cognitive decline of an individual during old age? And, if so, what would happen when this decline is accompanied by a neurodegenerative disease? The next section is devoted to these two questions.

COGNITIVE DECLINE AND BILINGUALISM

Watching the Spanish actor Pepe Rubianes in the theatre or on TV used to entertain me a lot, and it still does. Despite his inappropriate comments, his stories were fascinating, his explanations were very original, and, however exaggerated they were, I continue to wonder whether the stories were real or invented. In one of his stories, Rubianes pokes fun at people who go to great lengths their whole life to have a pension plan. He sarcastically said that when they turned eighty years old, they could start going wild, going out to clubs all night, eating at the best restaurants and travelling.

Regardless of whether we can save money or whether we even get to eighty, many of us worry about our financial situation and our retirement plans. Can we also save our cognitive ability as we grow old? In other words: is there something throughout life that protects us, even if only a little, from the cognitive effects of brain deterioration associated with healthy ageing?

There are ways to encourage some cognitive saving. Like most organs, our brain changes with age, and as we get older these changes have increasingly clear effects on our capabilities. Our brain not only becomes less plastic – which may be why it's more difficult for us to learn new things the older we get – but there are also some areas of the brain that become smaller or shrink, so as time goes by the brain loses volume. This shrinkage affects both grey and white matter. Of course, not all areas are affected equally, but in general, ageing effects are noted in many of them. Ageing is accompanied by cognitive decline that negatively affects many basic cognitive processes, such as attention, language, memory, and so on. Let's face it, it's not a very encouraging outlook.

Although the cognitive decline associated with age is inevitable, it seems that there are factors that can affect its progression and severity. To put it in another way, there are people who age better than others, cognitively speaking. Certain types of physical exercise and diet seem to have positive effects. In addition, other social or cognitive factors also seem to have a consequence for what we call *cognitive reserve*. In fact, it is estimated that around 30 per cent of people whose brain autopsies showed signs of suffering from Alzheimer's disease do not display the cognitive impairment that is typical of the damage observed. Why? Because of cognitive reserve.

Cognitive reserve is an easy concept to understand. Imagine two elderly people with the same brain deterioration, biologically speaking (the same reduction in white and grey matter volume in the same areas of the brain). If cognitive decline were determined only by such deterioration, then the two people should have the same problems. Well, it turns out not to be the case, or at least not necessarily. The same degree of brain atrophy may cause cognitive deficits in one person but not in another, or rather, not yet anyway. This would indicate that the latter person has a greater cognitive reserve than the former. The concept of cognitive reserve has caused some controversy but, thanks to epidemiological studies over the last decade or so, it is currently an accepted notion, although we do not fully know how it works. What we can say is that educational level and leading a rich and stimulating social and intellectual life seem to be beneficial in maintaining cognitive reserve. But perhaps that's not too surprising.

Before continuing, let me clarify one thing. Having more cognitive reserve does not mean that our brain will not deteriorate as a consequence of getting older. Nor does it imply that it protects us from developing diseases such as Alzheimer's or other types of neurodegenerative diseases. What it does mean is simply that the cognitive consequences of brain deterioration, whether normal or pathological, can be less harmful at the behavioural level. Of course, those consequences will depend also on the degree of cerebral involvement. And, in fact, as we can see in Figure 4.3, cognitive reserve has its drawbacks too. Although the greater the cognitive reserve, the later the symptoms of a neurodegenerative disease will appear, when these symptoms eventually manifest, cognitive decline becomes faster and marked. To put it simply: when you begin to realize that you are losing cognitive abilities, the progression of that loss gets faster. This happens because when cerebral neuropathology is relatively mild, symptoms may be lessened by cognitive reserve, but there comes a time in which the degeneration is so great that cognitive reserve can no longer help.

The first study that addressed the effect of bilingualism on cognitive reserve was performed in a hospital in Toronto. The experiment was simple. The authors examined the reports of 184 patients. All of them had been clinically evaluated and they met the criteria that indicated the probable presence of a neurodegenerative disease (Alzheimer's or another type of dementia). In addition, half of the patients were bilingual and the other half were monolingual. These two groups were matched in years of education, in cognitive performance (measured by a standardized neuropsychological test), and in occupational status. The researchers asked patients two important things: first, how old they were when they first visited the neurologist; second, how old they were when the signs of cognitive deterioration first appeared. This data was collected during the first visit to the neurologist, when he asked the patient and relatives for how long they had noticed the symptoms. The results were surprising and spectacular. The bilingual patients had first seen a neurologist three years

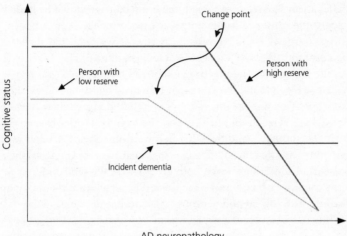

Figure 4.3. Representation of the cognitive decline associated with the neuropathology of Alzheimer's disease. The thick line represents a person with high cognitive reserve and the thin line shows an individual with low cognitive reserve. With moderate levels of neuropathology, the cognitive capacity of the person with less cognitive reserve will begin to decrease and show signs of Alzheimer's sooner.

later than monolingual patients. And this was not because they resisted seeing a doctor. In fact, monolinguals reported having noticed the symptoms at a younger age (seventy-one years) than the bilinguals (seventy-five years). These dates suggest that bilingualism helps the development of cognitive reserve, which in turn reduces the negative effects of brain deterioration. And the consequences of the effect, in this case, are not small. The difference is . . . four years! You can imagine that these results were quickly picked up by the media and publicized to great fanfare.

The phenomenon of cognitive reserve can also be evaluated in other ways. One of them is by analysing and comparing the degree of neuropathology and the degree of cognitive impairment. If patient A and patient B show the same cognitive performance, and we assume that A has a greater cognitive reserve than B, then we could predict that patient A should have a higher degree of neuropathology than patient B. And even though the brain damage would be higher in patient A, this would not imply worse performance because patient A has a greater cognitive reserve. It is as if we predicted that the famous football player Lionel Messi will continue to play fairly well despite the fact that he is getting older, and his legs aren't as strong; he has so much football reserve that he will go on for many years, or so I hope. This was the line of thought followed in a study that evaluated the cerebral atrophy of forty patients diagnosed with Alzheimer's. Half were bilingual and the other half were monolingual. What was important was that, in addition to being equal on other variables, such as age and educational level, both groups were the same in terms of their cognitive performance as measured by standardized tests often used in neurological and neuropsychological evaluations. What did we find out when brain atrophy was measured for these two groups? It turns out that the bilingual speakers showed greater atrophy than the monolingual speakers. This atrophy was not present in all areas of the brain, but only in those areas that are generally measured to distinguish patients with and without Alzheimer's disease. As such, although bilingual individuals showed greater neuropathology than monolinguals, cognitive impairment was similar in both groups, presumably thanks to the greater cognitive reserve of bilinguals.

But is bilingualism really the underlying factor explaining the difference between these two groups? Could we not attribute that difference to other variables that may be related to being bilingual? These studies were conducted in Toronto, one of the most multilingual cities in the world, which seems like a good place to carry out this type of work. However, this opportunity comes with an associated cost. A large number of bilingual people from this area have a history of immigration. In fact, according to the United Nations, Toronto has the second-highest percentage of foreign population in the world. Therefore, many of the people in the studies are likely to be immigrants or from immigrant families. This is important, because one might think that the difference between bilinguals and monolinguals could be due to ethnic differences and lifestyle habits (diet, for example), and not necessarily due to their linguistic status. In addition, and to further complicate things, some studies have shown better cognitive performance among immigrant children compared with that of non-immigrant children. So what do we do now? How can we know which factor is enhancing cognitive reserve?

One possibility is to carry out similar studies in which the two groups of speakers do not have a history linked to immigration. One of the places where we find this situation is in the city of Hyderabad, in the south of India, where several languages have been spoken for many centuries, and where around 60 per cent of the population is able to speak at least two of them. In 2013 researchers from Nizam's Institute of Medical Sciences in Hyderabad examined the clinical records of 648 patients diagnosed with dementia, 391 of whom were bilingual. The results were surprisingly similar to the findings in the Toronto study. Bilingualism delayed the onset of symptoms of dementia by about four years. And this study could also analyse the potential effect of education, given that in this region there are many people who are illiterate and have never been to school. In fact, in this sub-sample of illiterate individuals, the effect of bilingualism was even greater, delaying the symptoms of dementia for about six years.

This all sounds very promising, but I hope you have realized that these studies could have the same problem that we saw in the previous chapter, when we compared the impact of bilingualism on certain cerebral structures. Yes, the chicken or the egg problem. What if it turns out that it is not bilingualism that implies greater cognitive reserve, but the other way around? That is, is it possible that those people who show enhanced cognitive abilities are not only more prepared to learn two (or more) languages, but they also show a greater cognitive reserve when they are older? If that were so, the story we would be left with would be very different. As Rod Stewart sang, 'some guys have all the luck'.

How can we find out exactly what is happening? The ideal situation would be to know the cognitive abilities of each person before they start learning a second language. So we would have a baseline that would allow us to establish a cognitive capacity benchmark for monolinguals and those who will become bilingual. Knowing that, we could then compare their cerebral deterioration and attribute the variations to language learning and not to pre-existing differences between groups. If we started the study from the very beginning, it would take us around seventy years to answer the question, since we would have to wait for the individuals we analyse to show symptoms of dementia. It seems to me that there are not many scientists willing to carry out such a study. But sometimes, just sometimes, luck smiles down on us.

Take a look at the interesting study that took place in Scotland. It turns out that in June 1947 all Scottish children born in 1936 had their intellectual capacity measured by researchers at the University of Edinburgh. Yes, you read that correctly, this study analysed *all* (or almost all) eleven-year-old children born in Scotland (about 71,000 of them). This was called the *Lothian Birth Cohort* and today researchers are still evaluating the cognitive capacity of these people, who are now in their eighties. So it seems, then, that we *do* have a baseline. Further, as would be expected, given the social characteristics of Scotland, most of these children were monolingual. And because the cognitive performance of these same individuals was reassessed when they were seventy-three years old, we also have something with which to compare the baseline. So we have results from two different developmental points, at ages eleven and seventy-three. The question then is how these two results are related.

The first finding was not particularly surprising: test scores that the participants obtained when they were eleven predicted their cognitive performance quite well at seventy-three, suggesting that intelligence is a fairly stable feature. Think of it this way: if a child is much taller than average when he is eleven years old, he will likely continue being so when he is in his seventies.

What is more interesting to us is that some people performed better at seventy-three years old than would have been expected, given their performance in childhood. That is, the cognitive decline in their seventies was smaller than their tests at eleven years old would have led us to believe. These people had done something during their lives that reduced the expected impact of ageing on cognitive performance. Well, it turns out that the individuals who had learned a language after age eleven (262 of the 853 participants; those who already knew two languages before were excluded from the analysis) showed better cognitive performance than expected. From my point of view, so far, this is the most convincing evidence suggesting that bilingualism can lead to the development of greater cognitive reserve. If you know of similar databases outside Scotland, let me know.

I would love to be able to finish this section here, and conclude that bilingualism helps the development of cognitive reserve, which protects individuals from the consequences of brain damage, at least for a few years. It's not that I want to compete against monolinguals, but that most of the people I know are bilingual and, therefore, I would like it to be true that my friends have better cognitive reserve. But, as often happens, the issue is not so clear. These results attracted so much attention that numerous laboratories and hospitals went to work and started to look at the medical records of different groups of patients to see whether the effect of bilingualism on cognitive reserve could be detected from their records. The results are a bit contradictory and seem to have muddied the waters. Some studies find an effect of bilingualism and others do not, and the worst thing is that we still do not know what variables determine whether such an effect is found. Let's look at some examples.

One of the largest studies to date evaluated the cognitive decline of 1,067 people of Hispanic origin who lived in north Manhattan. The sample included monolinguals of Spanish and Spanish-English bilinguals and their cognitive performance was evaluated for twenty-three years, starting from the 1990s. Every two years they were subjected to cognitive tests that measured the cognitive deterioration associated with ageing. This study was allowed to observe how bilingualism affected an individual's development over time. The first conclusion was that the higher levels of bilingualism were associated with better cognitive performance at the beginning of the study, that is, twenty-three years earlier. But this doesn't mean much (remember, the chicken or the egg). However, cognitive decline did not depend on the level of bilingualism. That is, bilingualism did not seem to provide more cognitive reserve. Another interesting result has to do with the likelihood of developing dementia. In this case, the results were also negative, and those people who were bilingual did not have lower odds of developing a neurodegenerative disease.

Other studies have shown more complex patterns. For example, a study conducted in Montreal noted that bilingualism did not delay the symptoms associated with dementia. However, this delay was present in individuals who knew more than two languages. Also, some studies looking at older people have observed that bilingualism seems to promote the development of greater cognitive reserve, but only in those in relatively low socioeconomic situations.

This is the state of the field today. We still can't say for sure whether bilingualism protects us better from cognitive deterioration or not. I can tell you my opinion, though. From my point of view, there is enough experimental evidence that suggests that there is an effect of bilingualism on cognitive deterioration. However, we are far from understanding what conditions facilitate this effect and make it more intense and detectable. It is very possible that bilingualism interacts with many other variables, such as socioeconomic status and educational level. Bilingualism may not have substantial effects on every person and, perhaps for this reason, it is often difficult to detect such effects. As the song goes: 'Life is just like that, I have not invented it.' But don't throw in the towel just yet; you know that patience is science's best friend.

5

Making Decisions

Sometimes I feel that prizes are awarded too hastily. Take the Oscars, for instance. I don't understand the criteria used for nominating and awarding these honours, especially when presented by Warren Beatty . . . but that worries me very little, because it only has to do with Hollywood. It irritates me more when I do not understand Nobel prizes and, especially, the Nobel Peace Prize. How can there be so many disparities between the reasons for which a Nobel Peace Prize is awarded to two different people? That leads to paradoxical situations like when such an award is given to, for example, Henry Kissinger and Nelson Mandela. Incomprehensible. The political and philanthropical careers of these two people could not be more different. Whereas Kissinger is still accused of numerous violations of human rights around the world, Mandela is revered as the champion of the anti-racist struggles in South Africa. While the law may pursue Kissinger, the world admires Mandela. But let's leave Kissinger aside and stick with Mandela for now.

Nelson Mandela spent twenty-seven years in prison as a result of his struggle against South Africa's apartheid system. During that time, among many other things, he devoted himself to studying Afrikaans, which was the language used by the descendants of the Dutch settlers who established a segregated regime in South Africa for more than forty years. What motivated Mandela to want to learn that language? In fact, most of the South African population spoke Mandela's own mother tongue, Xhosa, a Bantu language. For obvious reasons, a good part of this majority strongly resented the people who spoke Afrikaans, since they identified it with the enemies who segregated and oppressed them. Some say that Mandela wanted to learn Afrikaans to be treated better by the prison guards. Others believed, and this seems to be what Mandela himself thought, that the best way to prepare to fight against enemies was to know their customs, their tastes, and their *language*. Be that as it may, Mandela once said something that resonates with the topic we'll explore in this chapter: 'If you talk to a man in a language he understands, that goes to his head. If you talk to him in *his* language, that goes to his heart.' Maybe that's what Mandela had in mind when he started learning the enemy's language. He wanted to be able to communicate with them in a language that would speak to their heart and not just their mind. In this chapter, we will see how right Mandela was, and how our emotions and processes that we put into play when we make decisions can vary depending on the language we use in each moment.

THE COMMUNICATIVE CONTEXT
IS EVERYTHING

In Chapter 1 we described some of the challenges faced by babies who are exposed to two languages. These children will learn the two languages in very similar social contexts. Let's say one parent speaks Dutch and the other parent speaks Swedish, so the situations in which the two languages will be used are quite similar. That is, both will praise and scold him, but in two different languages. However, a large part of the population learns a second or third language in very different social contexts from those in which they learned the first. Many times, this learning is limited to academic situations, in which the vocabulary and grammatical properties of the language are acquired in a formal way. There are many differences between being raised bilingual from the very beginning and learning a second language in an academic setting. Here we will focus on one key difference, the social use of language, which will lead us to look at the part played by emotions.

When a language is learned in an academic setting, its social use is often quite limited. This can often make students ask themselves about the usefulness of learning a second language. In my opinion, the learning context can affect various levels of processing differently. A context limited to the academic environment or, if you will, detached from the social use of language, *could* not be that important in terms of how the grammar and sound properties of that language are acquired. However, it may be important for what we call 'pragmatics', or language use. In very general terms, pragmatics has to do with how a context or situation affects the interpretation of words or communicative acts. It has to do with the inferences we make about what a speaker communicates to us, beyond what they are simply saying; which language registers are appropriate for certain communicative situations and so on.

For example, when we talk ironically or sarcastically, we say the opposite of what we want the listener to hear. At a restaurant, I usually ask if their special of the day is good or very good. If I just asked whether it is good, I might be questioning the quality of the establishment, and the waiter would be forced to say 'of course it's good', and maybe even in an apathetic tone. By my asking whether it's 'good or very good', the waiter understands what I am trying to find out without feeling insulted. You would be surprised to know how many times they respond with a simple 'it's good', which then makes me end up ordering something else. Pragmatics is also involved when we use language indirectly, as when Vito Corleone in *The Godfather* said 'I'm gonna make him an offer he can't refuse', but what he really meant was 'he will accept my offer, yes or yes'. For this type of communication to be effective, the interlocutors must have a similar understanding of the context in which the communicative act is performed. It's not the same when your boss tells you 'I'll make you an offer you can't refuse' as when your child says it. Not only does context matter, it is also necessary for speakers to have sophisticated knowledge of word meaning, including how those meanings may vary depending on the context.

This is, perhaps, the most difficult thing we face when learning a second language. In fact, it is difficult even in our first language, as we can see with children who usually interpret language literally until approximately six years old. I won't present here the different theories about pragmatics, many of which come from the philosophy of language, but instead I will focus on a few examples that might be familiar if you have learned a foreign language. These examples will set the stage for the following sections in which we will focus on the relationship between emotions, the processing of a first language and second language, and decision-making.

'I don't get it', says the little voice in our head when someone in a group of friends tells a joke in a foreign language. Everyone laughs, but you are lost. You manage a smile, maybe infected by everyone else's laughter, or maybe because you don't want to show that you didn't understand what the joke was about. But you didn't understand, and you know it. You become depressed and curse the eight years of classes in that foreign language back in school. It's not that you don't understand words or phrases, you get those perfectly, but you don't find anything particularly funny about them. Why is it so difficult to understand humour in a foreign language?

Humour is an extremely complex communicative act in which many different tricks come into play. We can make jokes that involve irony, indirect language, surprise, tone of voice, double meaning of words, phonological similarity, and so on. But above all, humour often requires one to forget literal meaning and realize that what is being said isn't necessarily what should be interpreted. This contrast causes laughter. How do you learn something like that? How is that taught? Can it be taught in school? Very doubtful. Language classes tend to focus on the literal meaning of words and not that much on how to use them in various situations. This may only be learned with language use in a social context, that is, interacting with other people in situations in which the meaning of words depends on what is intended. To me, one of the milestones in acquiring a foreign language happens when, for the first time, we are capable of explaining a joke in that language . . . as long as others laugh too, of course.

Another example is that of 'bad' words and expressions. I won't list the many swear words that are out there but you know the kinds of words I am talking about. Surprisingly, swear words in a foreign language are learned relatively quickly, maybe because they seem fun to learn. We all use swear words, to a greater or lesser extent. There are different explanations as to why we do it and if you're interested you should read Steven Pinker's reflections on this subject in his wonderful book *The Stuff of Thought*. From my point of view, there's not a problem with using swear words; the problem is knowing when to use them and, above all, which ones to use in specific communicative situations. And that is difficult to learn or discover, because it depends to a large extent on the social, linguistic, and emotional context. The meaning of those words is oftentimes quite similar and, in fact, they can take different meanings depending on the context. Choosing one or another word goes, to a large extent, hand in hand with the emotional intensity associated with each word and the communicative situation. No one wants to go too far or fall too short. Furthermore, it is common for these words to come out of our mouth almost automatically without their doing much more than making the message more emotional to get someone else's attention. Think how often you use a swear word when something unexpected happens to you . . . and no one else is around.

These words and expressions seem to have a fundamental emotional component that shapes their meaning and guides their usage in each context. How do we learn this when the foreign language is acquired in a classroom? Maybe it's impossible. And if you do not agree, try explaining to someone when it's more acceptable to say *sh*t* in some contexts and to say *f*ck* in other situations. Good luck. It's not as if we don't know the translations; these words are the first ones that students look up, in the brand-new dictionary that their parents have bought them, when the course begins. It is true that for some swear words there are no translations, since for cultural reasons they are only used in one language, but because many have to do with sexual themes or bodily functions, we have many versions. But this still isn't enough, since what we do not know is how to use them, and to a large extent we do not understand them correctly even when we hear them. This often happens not only because of lack of pragmatic knowledge in the foreign language, but also because of the fact that those words do not seem to *sound* like they should in their corresponding translation. But what does it mean that they don't *sound right*? Well, the emotional reaction that we experience when we hear or say them doesn't seem to be the same, and because to a large extent their meaning is affected, they do not sound like they should. The result of all this is that we not only swear wrong in a foreign language, but we don't seem to care too much. There seems to be so much emotional distance between us and a foreign language that it allows us to use swear words more easily than in our first language, which, believe me, surprises the native speakers of that second language and makes them uncomfortable. So, another milestone in acquiring a foreign language would be to know how to use swear words and expressions appropriately. And beyond that, if they *sound* bad to us, they will make us *feel* bad.

I'm afraid that some milestones like explaining jokes and using swear words correctly in a second language will not be taken into account by specialists when designing teaching methods. In fact, they may be irritated by the previous paragraphs. After all, you know the most important thing is that the students learn their irregular verbs and not make too many spelling mistakes. Maybe this is not the most important thing, but rather the most comfortable to teach in an academic context. That's just how it goes. But it doesn't matter too much if students don't pay much attention in class because it seems that a lot of language learning is not acquired in academic contexts but rather by doing things such as playing PlayStation online, listening to music, or watching YouTube videos; that is, using the language in a social context.

LANGUAGE AND EMOTION, OR WHEN WORDS DON'T EXPRESS WHAT THEY SHOULD

Nelson Mandela thought that by speaking to someone in his first language, the message would go to his heart and, in the second, would go to the head. A bit exaggerated, perhaps. Many centuries earlier, Charlemagne, who spoke French and Latin, and understood classical Greek, said that 'knowing another language is like having a second soul'. Even more exaggerated. But is there any truth to this? To what extent are our emotional reactions different in the first and second language?

Let me give you an example because I think it sums up well what comes next. Although I've been speaking Catalan and Spanish since I was very young, my mother tongue, the language in which I feel most comfortable, is Spanish. But I do usually communicate with my son in Catalan and also use it very often with friends, colleagues, and students. When my son was about eleven years old, we were having an argument – who knows about what anymore – and things were getting tense. When he went too far, I stopped speaking Catalan and switched to Spanish. Up until then, everything had been more or less normal. The funny thing was my son's reaction. He said: 'No, Dad, not in Spanish!' 'What? What did you say?' I asked. He replied that he had noticed on other occasions that, when I got angry and switched to Spanish, it was because I was really irritated. It was as if my son had learned that the emotional intensity with which I could express myself in my dominant language was greater than in the non-dominant one, and that when I was really angry I would pull out the Spanish. I laughed and my anger passed. He was grounded anyway.

The studies that have explored this issue have taken two different approaches. First of all, we have field studies of bilinguals who are asked about their emotions and feelings when using their two languages. Work by Aneta Pavlenko and Jean-Marc Dewaele, among others, suggests that our perception of the emotions we experience when speaking our mother tongue is much greater than that when speaking a foreign language. In other words, it doesn't *sound* the same. These studies are very useful because they explicitly ask people about their relationship with the languages they use. However, they may also be affected by response biases and by judgements about how people think they feel in different languages. Put another way, if someone asks me whether I feel the same when I hear 'I love you' as when I hear '*Te quiero*', I am likely to say that I feel more emotion in Spanish, my first language, although it is not immediately obvious that the feeling is the same as the reality. One thing is what I *think* I feel, and another is what I *really feel*. More studies are needed to untangle this.

The other approach has been to carry out more indirect experiments to see to what extent the emotional reaction to words or phrases in both languages is different. The results here are more inconsistent. Let's look at a few of these studies.

In previous chapters, we mentioned the Stroop effect when we talked about attentional control. This happens when an irrelevant aspect of the stimulus interferes with the participant's task at hand. For example, imagine that you are asked to identify the font colour of a word. In principle, the meaning of the word shown is irrelevant, you just have to name its font colour. Therefore, it should take the same amount of time to say that the word *motorcycle* is written in black as it takes to say that the word *red* is also in black. You see the trick, right? Answers are slower when the meaning of the word also corresponds to a colour than when it does not. In other words: an irrelevant dimension of a stimulus (the meaning of the word) interferes with the task that must be done (name the colour of the font) and consequently affects our performance. Well, it turns out that the same type of study can be carried out with emotion words. What has been found is that it takes longer to say the font colour of a word when its meaning provokes an emotional reaction (*love* or *death*) than when it does not (*table* or *beer*). These results suggest that the emotional value of words captures attention automatically, and when emotions are high, it causes us to get distracted and have fewer resources to allocate to the task at hand: naming the font colour. In fact, there are many studies that have shown that this effect is reduced when the words are presented in the participants' second language. That is, the emotional value of words in the foreign language seems to be smaller and, consequently, grabs our attention less and doesn't interfere as much. Nevertheless, other studies have been less successful in showing a difference between languages in these types of experiments. So the question is still unanswered.

Other studies have investigated the psychophysiological reaction that emotion words cause as a consequence of changes in the autonomic nervous system. How can we measure these reactions? There are certain indicators of changes brought about by emotional situations, such as levels of electrical conductivity of the skin, heart rhythm, or pupil dilation. When we are in an emotional situation, our skin conductance increases because of sweating, our pulse goes up, and our pupils dilate. In a series of studies conducted by Catherine Caldwell-Harris at Boston University, it was observed that the electrical conductivity response or electrodermal response to emotional stimuli is smaller in a second language that has been learned after childhood. In one of these experiments, which is particularly curious, phrases corresponding to reprimands were used in the participants' native language or in the foreign one, like for example,

Aren't you ashamed? These phrases caused a greater change in electrodermal response than neutral ones such as *The car is blue*, but only in the participants' first language.

These results show that social experiences from our childhood can have effects on language processing when we are adults. It's as if an automatic association were created between the expressions that our parents used to say to us and the emotional states that these expressions provoked in us. This association will always exist in the language in which our parents speak, and not so much in languages that we have learned as adults and in academic contexts.

There are also studies that have explored the brain activity associated with emotional messages presented in a first and second language. For example, in a study carried out at the Free University of Berlin, the brain activity of German-English bilinguals was evaluated while reading passages of *Harry Potter* with neutral or positive emotional implications. The results were clear: emotional fragments activated areas of the brain related to emotional processing, such as the amygdala, to a greater extent than neutral passages. However, this only happened when the participants read the texts in their first language (German). In their second language (English), there were fewer differences in brain activity for both texts.

Although these studies point to a lesser emotional response in a language learned after childhood, there is still a lot of work to be done. For example, we do not know whether such reduction is caused by the fact that it involves a second language or because of the social context in which it was learned. And we don't know if this has *only* to do with age of acquisition. My sense of it is that all these variables contribute to our emotional response, although perhaps the most determining one is the social use of language.

DECISION-MAKING: INTUITION AND REASON

I started my research in the Department of Psychology at the University of Barcelona more than twenty-five years ago. It was the summer of 1991 and I had just finished my second psychology course. I went to the Department to see whether there was anything I could do over the summer. My interest was linked to cognition in general, but especially to decision-making and problem-solving. On the way, I bumped into Professor Núria Sebastián, who skilfully redirected my interests towards language and bilingualism. Goodbye, decision-making and problem-solving; hello, language. More than two decades later, I have achieved two things. The first is that I have continued collaborating with Núria. The second is, in the end, I've been able to investigate decision-making, albeit combined with bilingualism. As they say, 'where there's a will, there's a way'. This section is dedicated to decision-making and we should introduce some basic concepts about it before moving on to the subject of how our decisions can be affected by the language that we use.

Two of the most influential researchers of the last forty years in cognitive psychology have been Daniel Kahneman and Amos Tversky. Thanks to their work, we have learned a lot about the cognitive mechanisms that come into play when people make decisions. The fruit of these discoveries produced a new discipline that straddles cognitive psychology and economics, which has come to be known as behavioural economics. Both men deserved the Nobel Prize in Economics (or psychology, if it existed), although unfortunately, it took until 2002 and Kahneman was the only one who was around to accept it as Tversky had passed away. (Michael Lewis has written about their relationship in *The Undoing Project*.)

Their fundamental contribution was to develop a theory that had been proposed by Herbert Simon, who won the Nobel Prize in Economics in 1978. In short, the idea is that in a complex situation that requires a decision, people tend to simplify the details that the situation entails and take heuristic shortcuts instead of calculating the real probabilities of the options faced. This simplification and these shortcuts offer us intuitive solutions to the problem at hand. It is as if we saw the clear solution suddenly without having to consider all of the variables that are involved in the problem. These shortcuts often work and the intuitive solution to the problem is the one that best serves our purposes. For example, if you are thinking about ending a relationship and you start making a list of the pros and cons of ending or maintaining the relationship, consider it finished. Love does not work like that. On some of these occasions, we are almost unable to detail the steps we have followed to get to that solution; we do not know why we chose it, but it has worked. Part of that intuition comes from experience that we have accumulated more or less implicitly from when we faced similar situations. That implicit learning allows us to come up with a solution to the problem almost immediately in similar situations.

However, at other times these heuristic shortcuts involve a certain distortion of reality and of the probabilities of options that are presented to us. Depending on the context, these distortions can lead to somewhat irrational behaviour and decisions that are not optimal for our interests. We call them *thought biases*. If we always deliberately reasoned about the different variables of a problem, maximizing the expected value of our actions, then we would be behaving like *Homo economicus*, just like some classical thinkers of economics had thought. But it turns out that we are *Homo sapiens*, and our decisions are affected as much by intuitive processes. As Tversky said regarding his line of research: 'My colleagues study artificial intelligence, I study natural stupidity.'

Let's take a classic example that comes from Tversky and Kahneman:

> Linda is 31 years old, single, outspoken, and very bright. She majored in philosophy. As a student, she was deeply concerned with issues of discrimination and social justice, and also participated in anti-nuclear demonstrations. Which is more probable?
>
> a) Linda is a bank teller.
> b) Linda is a bank teller and is active in the feminist movement.

I would venture to say that a good number of readers chose the second option. You must realize, at least, that many had their doubts. The answer is extremely obvious if one stops to think: the first option is the most likely. Why? Because the probability of two things happening together cannot be greater than the probability that only one of them will occur. Simply put, if Linda is both a bank teller and an activist, then she must be a bank teller; while it is possible that she is a bank teller but not an activist. In the original study, around 85 per cent of the respondents chose the second option, falling victim to the *conjunction fallacy*. This error seems to arise from what is called the *representative heuristic*, whereby the second option agrees more with the description that was given about Linda, although it is clearly less probable in logical terms. To put it another way: given the preamble, it makes all the sense in the world to think that Linda is a cashier and an activist, although that is actually less likely. But if we think about it more carefully, we can find the right answer without any problem.

These types of studies, among others that we will look at later, have led researchers to postulate that there are two systems at play in the decision-making process. On the one hand, an intuitive system that sets in motion heuristic shortcuts and provides us with solutions quickly and almost automatically – or what in technical terms has been called 'System 1 decision-making'. This is the one that makes us see things clearly (Linda is a bank teller and an activist). On the other hand, we have another logical and reflexive system called 'System 2', which allows us to consider the different variables of the problem and reach conclusions that go beyond those proposed by our intuition. This system is the deliberative one. However, this system is slow, cognitively demanding, and costly in terms of mental resources; that is, we have to stop and think. Our decisions are influenced by these two systems in complex ways and, in fact, the interaction between the two is what will ultimately guide us. Next we will analyse to what extent the language in which a problem is expressed can be one of the factors that affects our judgements, preferences, and decisions.

BE CAREFUL WHAT LANGUAGE YOU USE: IT MAY AFFECT YOUR DECISION-MAKING

One of the factors that increase the contribution of intuitive processes to our decision-making is the emotional reaction that a particular situation provokes. In very emotional situations we let ourselves be guided more by intuition or, if you will, it's harder for us to stop, think, and reason about what we have in front of us. Software developers have tried to remedy these impulses with programs that delay sending emails for a few minutes or even hours, and that allow you to undo something that has already been sent. These programs let us think twice, something that can prevent us from getting into trouble. To say it in a somewhat simpler way: the less emotional the situation, the better our control of intuitive processes and the greater our efficiency controlling the biases that heuristics can bring.

In previous sections we reviewed some evidence which suggested that the use of a second language learned as an adult, or in an academic context without (or with little) social use, could imply a reduction in the emotional response in that language. Neither swear words, reprimands, nor the spells of Harry Potter sound the same in our native language as they do in other languages. And this *not sounding the same* also makes us less emotional.

Here is where this leads us: if thinking biases associated with heuristics are facilitated in conditions of emotional intensity, and the use of a foreign language reduces the emotion caused by the message, then those biases will have less influence when we make decisions in the second language. If this were the case, the decision-making in this context might respond more to deliberative and logical criteria than in a situation involving the native language. System 2 would have more chances of taking the reins.

The first study that analysed this question by exploring the 'framing effect' in decision-making was directed by Boaz Keysar at the University of Chicago and published in *Psychological Science* in 2012. This framing effect refers to the fact that our decisions can change depending on how the options are presented when facing a particular problem. This example from the study lets you judge for yourself:

Gain-frame

Recently, a dangerous new disease has been going around. Without medicine, 600,000 people will die from it. In order to save these people, two types of medicine are being made.

If you choose Medicine A, 200,000 people will be saved.

If you choose Medicine B, there is a 33.3 per cent chance that 600,000 people will be saved and a 66.6 per cent chance that no one will be saved.

Which medicine do you choose?

What medicine have you chosen? Do not worry, one is not a better option than the other. In fact, the expected value, in economics, is the same for both. The only difference is that the result of choosing medicine A is safe in the sense that we know what will happen, whereas choosing medicine B is a gamble. The choice depends on what is called *risk aversion* or, in other words, it depends on how daring an individual is. In any case, we know that about 75 per cent of people choose option A, in which 200,000 people are saved, even though it also means the certain death of another 400,000. As the saying goes, 'a bird in the hand is worth two in the bush'. Up until now, everything seems logical. But now comes the trick and the pioneering discovery. Other participants were presented with the same problem and options as in the gain-frame version but with a different focus:

Loss-frame

Recently, a dangerous new disease has been going around. Without medicine, 600,000 people will die from it. In order to save these people, two types of medicine are being made.

If you choose medicine A, 400,000 people will die.

If you choose medicine B, there is a 33.3 per cent chance that no one will die, and a 66.6 per cent chance that 600,000 people will die.

Which medicine do you choose?

Hmm. Did you change your decision? Would you now take more risk and choose medicine B? At a minimum, the safe option doesn't seem so attractive now that the emphasis is placed on lives that will be lost (loss-frame version) and not on lives that will be saved (gain-frame version). Both problems are identical in terms of their consequences and, therefore, the decisions, whatever they may be, should be the same in both cases. That is, if we were *Homo economicus* ... but it turns out that we aren't. In the second problem, the number of people who give the riskier answer (medicine B) is much greater than in the first problem. Why? Because in the first, the option in which the result is safe (medicine A) is presented in terms of gain (how many lives are saved) and in the second, it is described in terms of loss (how many lives are lost), and as humans, we hate to lose lives, money, or anything. We suffer from what is called *loss aversion*, and given that in the second problem we can see the number of people who will die, we risk more. It's much like the saying 'in for a penny, in for a pound'. The important thing about this effect is that our decisions are not only determined by the expected value (or consequences) of the options that are presented to us, but also by how these options are described and framed. If a safe option is presented in terms of profits, it is more likely that we choose it to avoid risk. If this same option is presented in terms of losses, we tend to take more risks than before.

Keysar's discovery is that this difference in answers, according to whether the alternatives are presented as gains or losses, disappears when the problems are stated in the participants' foreign language. It's as if when facing these decisions in a foreign language, the feeling of loss aversion doesn't affect us. Surprising, isn't it? One would like to think that our judgements, preferences, and decisions should be guided by a calculation of probabilities and a rational evaluation of the options that are offered to us. The irrelevant aspects should be ignored when we make a decision. In any event, if those irrelevant aspects, such as presenting the problem in terms of losses or gains, affect our decisions, they should do so independently of language. Well, that's not how it works: language *does* affect our judgements and preferences. Researchers have attributed this phenomenon to a reduction in emotion associated with a foreign language, which decreases loss aversion and causes participants to behave in a more consistent manner . . . more rationally, if you will. That is, the loss-frame version would not cause such a negative emotional effect in the foreign language and, therefore, would not lead to more risky responses compared to the gain-frame version.

When I first read these results, I couldn't believe it. Not only were these findings very surprising, but they also had very important social, economic, and political implications. How many people are making decisions about problems discussed in their second or third language? How come we are more logical and consistent in a language in which it is harder to communicate? Is this a good or bad thing? What language are our politicians speaking nationally and internationally? Should companies start conducting their meetings in a foreign language? 'Not so fast,' I thought. This discovery would only have practical consequences if the phenomenon were of a general nature, and not limited to the framing effect described above. So we got to work, and started to undertake a series of studies on the inter-action between decision-making and language. I had begun to study these topics in the Faculty of Psychology at the University of Bar-celona some twenty summers ago and I was now getting back to it at the Centre for Brain and Cognition at Pompeu Fabra University.

Our studies showed that the effect of foreign language in decision-making was indeed coherent and generalizable to other situations. I'll give you a couple of examples. In one of the experiments we ana-lysed risk aversion. As we've seen, risk aversion basically refers to the notion that human beings tend to prefer safe options over others that are more risky, despite the fact that the safer ones are not necessarily those that can benefit us more. In other words, the expected value of the option we choose is less than the expected value of the other option that is offered to us, but the first option is more secure and the second carries a risk. Imagine that I asked you to play one of two lotteries. In the first (lottery A), you have a 50 per cent chance of winning £2, and a 50 per cent chance of winning £1.60. So we flip a coin in the air; if it's heads, you win £2, and if it's tails, you win £1.60. In the other option (lottery B), if it's heads, you would win £3.85, and if it's tails you win £.10. I wish all gambling were like this and we got money regardless of whether we win or lose. But in any case, which one would you choose? In lottery A, the minimum you would win is £1.60, which is much more than the minimum that lot-tery B guarantees (£.10). Lottery A, therefore, is more secure in terms of your assured gains if things go wrong. However, if the result is positive for you and you win the game, the prize in lottery B is almost twice as much as lottery A's (£3.85 vs. £2).

Homo economicus would have no doubt and, after a relatively easy analysis, would see that the second lottery has a higher expected value and, therefore, would choose it. Yes, you read correctly, *Homo economicus* would choose to play lottery B, while you may have cho-sen to play lottery A. But why did you possibly make this choice? Because lottery B implies a greater risk than lottery A in the sense that it ensures less money if you lose. When we asked for this type of decision to be made in a foreign language, English in the case of our students in Barcelona, the participants tended to show less risk aversion. In other words, they choose the safer lottery (A), although it has less expected value, much less often when presented in their foreign language than in their native language. This all shows that facing problems in a foreign language leads to better decisions. We believe this has to do with the fact that the emotional reaction which triggers risk aversion is less intense when the problem is described in a foreign language. So now you know, if you have to choose between casinos, go to one where they don't speak to you in your native language . . .

Another example has to do with what is called *mental* (or *psychological*) *accounting*. This refers to how we categorize the value of our economic transactions. Let's look at a situation that might be familiar to you. You go out on a Saturday to buy a jacket that you've had your eye on in a nearby store for a couple of weeks. You don't really need the jacket, but it's the first purchase of this season and you think that your efforts at work over the last several months deserve a reward. You have no doubt about that and, after all, we all have the right to spoil ourselves from time to time. You bump into a friend just as you are getting to the store, and he tells you that another shopping centre has the same jacket for a cheaper price. The jacket costs £125 in the store next door and £120 in the shopping centre, but for that, you would have to get in your car and drive ten minutes to get there. What would you do? Would you go get the car and drive to the shopping centre to save £5? There is no right answer to this question. Your decision will depend on many factors, including how stingy you are and how much time you have to go shopping. But now imagine that, instead of going out to buy a jacket, you were going for a scarf that costs £15 and your friend told you that it was £10 at the shopping centre, that is, £5 less. Would you be more willing to take the car? The answer is probably yes, or at least it's more likely than in the case of the jacket. Strange, isn't it? Both in the case of the jacket and the scarf, the trip to the shopping centre implies saving £5. Isn't that the same? The point is: it does not *feel* the same.

In our study we explored the presence of a foreign-language effect in this phenomenon. To do so, we proposed the following situations to different groups of participants:

Discount on a £15 item

Imagine that you want to buy a jacket for £125 and a calculator for £15. The salesman tells you that the calculator you want to buy is on sale for £10 at their other shop, located a twenty-minute drive away. Would you make the trip to the other shop?

Discount on a £125 item

Imagine that you want to buy a jacket for £15 and a calculator for £125. The salesman tells you that the calculator you want to buy is on sale for £120 at their other shop, located a twenty-minute drive away. Would you make the trip to the other shop?

The two situations are equal in terms of the total cost (£140) and in the total purchase discount (£5). The only difference is that in one case the discount is on the lesser-value object (which costs £15), whereas in the other case the discount is on the object of greater value (which costs £125). The results were clear. When the text was presented in the participants' first language, about 40 per cent of them stated that they would go to the other store when the discount was made on the object of lower value, whereas only 10 per cent decided to accept the discount on the object of greater value. This difference was reduced by half when the problem was presented in their foreign language (English). It's as if hearing it in the foreign language had led to a more thoughtful decision in which, at the end of the day, saving £5 is saving £5 regardless of the value of the product and the percentage saved.

The effect of a foreign language on decision-making also seems to emerge in our evaluation of risk. For example, when some participants are asked to rate the benefits or risks associated with certain activities, it appears as though the risks are rated as less dangerous and the benefits more important when presented in the foreign language. If we ask a group of people how much risk they think is associated with nuclear power plants, they see it as less risky when the question is asked in their foreign language. So be careful when comparing surveys that are carried out in languages other than the participants' first language (think about, for example, job satisfaction surveys among migrant workers).

These results suggest that our decisions can be affected by the language in which the problem is presented. In fact, it would seem that in a foreign-language context we become more consistent and, if you will, more reflective than in our native language. But what is behind this effect? How does it originate? It is possible that when we face a problem in a foreign language, we do so with more caution and greater effort. This, perhaps, would reduce intuitive biases and help us to make more reasonable decisions. To put it another way, when the problem involves some difficulty from a linguistic point of view, we put on our thinking caps and rethink our decisions, which allows us to block intuitive responses (System 1) and to make us think twice about the options presented (System 2). Note that, according to this explanation, the effect of a foreign language is not so much related to the reduction of emotion as to the cognitive effort it causes. If this were true, we should find this effect in situations that don't involve emotional responses, which would have even broader implications for education and society in general.

We tested this hypothesis using the so-called *cognitive reflection test* developed by Professor Shane Frederick at Yale University. The version we used contained only three problems that were designed to provoke an intuitive answer in the participants ... which, in this case, was the incorrect one. Therefore, to answer correctly, the participant had to get rid of the reaction that comes to mind and think a little more carefully. In fact, some studies have shown that there is a correlation between how well participants do on this test and how well they do on general intelligence tests. Here are the problems:

1. A baseball bat and a baseball cost £1.10 in total. The bat costs £1 more than the ball. How much does the ball cost?
2. If it takes five machines five minutes to make five keyboards, how long would it take 100 machines to make 100 keyboards?
3. In a lake there is an area with flowers. Every day, the area doubles in size. If it takes forty-eight days for the area to cover the entire lake, how long would it take for the area to cover half of the lake?

Don't tell me that the answers 10, 100, and 24 didn't quickly come to mind. But I'm sorry to say that these are wrong. Those three numbers are the ones that come to mind quickly, almost as if we clearly saw them automatically, but they are the product of our intuitive system. In fact, the problems are designed to do so. If you think a little bit more, you will realize that the correct answers are 5, 5, and 47. If the effect of a foreign language involved greater cognitive effort, then it would be possible for the performance on this task to be better when problems arise in the foreign language. But this was not the case, and the success rate for these three problems was just as poor in one language as in the other. In addition, the tendency to give the intuitive response was equally frequent in both languages. It does not look like the foreign language has an effect on logical problems that do not involve the emotional system. But there is still quite a bit to explore in this regard.

WOULD YOU SACRIFICE ONE LIFE TO SAVE FIVE OTHERS?

For many of us, our beliefs and moral values are what define us as people. We do not identify ourselves as being tall, blond, rich, or strong, but as being more or less ethical, understanding, selfish, and so on. Unless we adhere to the words of Groucho Marx, 'Those are my principles, and if you don't like them . . . well I have others', we like to think that we have certain principles and moral rules and that these make up what we really are. We also believe, or want to believe, that our principles about what is right and wrong are relatively stable and do not depend on irrelevant things like time of day or weather. But is this true? Are these principles really that stable or, on the contrary, are they less consistent than we believe? Can they be affected by variables that have nothing to do with them? Here we will see how foreign-language use can modulate these principles.

There's a line of thought according to which some of our moral judgements about different situations are guided by emotional responses and not, or at least not necessarily, by a reflection of the appropriateness of the behaviour in the context in question. It is as if our intuition led us to a clear answer about what is right or wrong, without the need for more elaborate reasoning about the specifics of every moment. So sometimes we say things like: 'OK, this is wrong. Why? Well, because it's just wrong.' This answer that pops up in our mind would be guided, to some extent, by the same intuitive mechanisms that in the previous experiments allowed us to come up with immediate solutions to economic or logical problems through System 1. Some authors such as Jonathan Haidt and Joshua Greene have related this response to the moral rules proposed, for example, by Immanuel Kant and his idea of *deontology*. According to this, an action can only be judged as good or bad if the rationale for taking such action obeys a law that can be universal to or independent of people's interests or desires. As such, it has been argued that an intense emotional reaction would lead us to follow a moral rule automatically without thinking too much about its suitability for the specific situation.

Let's take an example. Consider the following moral dilemma, widely used to study our moral judgements and originally proposed by the American philosopher Judith Jarvis Thomson:

> A train is very quickly approaching five people. The train has a problem with its brakes and cannot stop unless a heavy object is put in its way. There is a very large man next to you. The only way you can stop the train is by pushing him onto the tracks, killing him but saving five people.

Would you push the man onto the tracks to save five lives? Probably not. In fact, we know that when faced with this dilemma, about 80 per cent of people choose not to push the man. It is possible that when you read this, you smiled or frowned and had an emotional, uneasy feeling. It's as if you automatically said: 'Nope, I wouldn't do that for sure.' You don't really know why but your response was probably fast, almost automatic . . . your emotional system decided for you. 'I wouldn't do that, and it's as simple as that'; you see it clearly – just as clearly as the baseball costing ten pence in the example above. Then the justifications and moral arguments come into play: whether a person's life is sacred and should never be used as a means to achieve an end; whether one can decide who lives and who dies; whether physical action against someone is acceptable or not; etc. Don't be fooled, the decision has already been taken and these arguments are only justifications to yourself. Your intuition has already helped you with the problem and provided a quick resolution: you would not push that poor man.

However, if we stop to think a little bit we will see that, from a utilitarian point of view, if we decided to push the man onto the tracks, we would save five lives. And isn't it better to sacrifice one life to save five? Well, to a certain extent that would depend on whether we take a moral utilitarian vision, in the sense that the expected result is maximized, or on the contrary, if we lean more towards a deontological vision, in which the moral law of not using a person's life as a means to achieve an end must universally prevail. This debate is, in effect, more complex, since on many occasions the definition of the term *utility* is somewhat difficult.

In any case, some researchers approach these moral judgements in the context of the two decision-making systems that we have described in the previous section, one that is a much more intuitive system that proposes answers almost automatically and one which allows us to evaluate the various options and their consequences in a more thoughtful way. My objective here is not to argue which decision-making system is right from the ethical or moral point of view. There is a lot of research on ethics and morality, and I'm not an expert on either. What does interest me, however, is to show you how our moral decisions can change depending on the context of the problem.

Consider the following dilemma devised by the British philosopher Philippa Foot, which in many ways is similar to the previous one:

A train is very quickly approaching five people. The train has a problem with its brakes and cannot stop. Five people will die if the train continues on its path. There is a way for you to divert the train, but that would result in a man dying.

Would you redirect the train to the other way? Probably yes. We also know that about 80 per cent of people would do it, sacrificing one person's life to save five other lives. That is, those people who had said that they would not push the man onto the tracks now say that if they are able to divert the train, they would sacrifice one person's life to save five. But aren't these two moral dilemmas the same? Yes, they are in terms of expected value from the action or inaction, that is, in terms of your decision's consequences (if you choose to act, one person dies and five are saved; if not, five die and one is saved). However, our emotional response to the two dilemmas is not the same, is it? The first is much more unpleasant than the second. And as the second one is less intense emotionally, we can stop to think and more calmly decide what sacrificing a life to save five is worth; that is, we use a utilitarian judgement.

And now comes the interesting hypothesis, which you might have guessed. If our emotional reaction to dilemmas presented in a foreign language is less than in the first language, then it may be the case that our moral judgements and decisions resulting from them would be affected by the language in which the dilemmas are presented. In other words, if we become colder (or less emotional) in a foreign-language context, then maybe we become more utilitarian. Is it possible that Groucho Marx was right after all and that our principles were much more easily influenced than we think?

Ask and you shall explore, right? We set out to test this hypothesis by presenting these two dilemmas to 400 native speakers of Spanish who spoke English as a foreign language. These participants were university students who had studied English in school for at least seven years. They did not use English in social environments, but they understood the text without problems. Half of the participants were presented with the dilemmas in Spanish and the other half in English. The results were striking. For the less emotionally charged dilemma, the one in which the train is diverted, the results were similar in both languages. Basically, 80 per cent of the participants opted for the utilitarian response, that is, to divert the train and save five lives. This result was expected given what we already knew from other studies. But what happened with the dilemma that supposedly causes greater emotional intensity? Would the decision to push the man onto the tracks depend on the language in which the dilemma was presented? Well, it turns out that it did; when the dilemma was given in the first language, only 17 per cent of the participants chose to sacrifice the man's life, while in English that option was chosen in 40 per cent of the cases. In other words, the percentage of utilitarian responses doubled when the dilemma was presented in the second language. So moral judgements do change according to language.

I realized that we had discovered something interesting when I was explaining these results to my mother and son over lunch and they both said at the same time: 'No way!' If people who were more than fifty years apart in age were surprised by the same phenomenon, it was because they could not believe that their moral judgements, what most identified them as individuals, could be affected by such an *insignificant* thing as the language in which a moral dilemma is presented. And believe me, my stories almost always bore them.

Before presenting the results to the scientific community, we decided to evaluate to what extent the change in decisions of the participants was affected by the fact of having used English as a foreign language and Spanish as a first language. You may have heard that some people think that English is the *language of business*, which could encourage more utilitarian views of the world (as if business were not conducted in Spanish, Chinese, or Russian). So we presented the same situations to native speakers of English who spoke Spanish as a foreign language. The results were the same. There were no differences between the languages when the choice was to divert the train, but there were when the choice involved pushing the man onto the tracks. In this case the participants tended to choose it twice as often in the foreign language than in the native language.

Even though these results have been replicated in several languages and laboratories, which suggests that they are robust and reliable findings, we still do not know their origin. I have presented them here following the hypothesis of the reduction of emotion in a foreign language, but there are other possible explanations. At the moment we do not have sufficient data to corroborate or refute them, but we should have soon.

FOREIGN LANGUAGE AS A SOCIAL MARKER

Not only can language modify our economic and moral decisions, it can also affect the way in which others see us. A second language also has an impact on how we make decisions about other people. As we saw in Chapter 1, children tend to use the language spoken by others to determine their social circles. Recall the study in which children were asked with whom they wanted to play and, among the candidates for being friends, there were children who were either native speakers of the same language, who spoke the language with a foreign accent, or who spoke another language. The children chose those who spoke their own language . . . and specifically those who spoke it without a foreign accent. In this section, I will show you that the effects of social categorization are also present in adulthood, both implicitly and explicitly. In addition, we will see that this categorization has effects on how others perceive us, and can be at the root of stereotypes and prejudices. Do you remember the steps that Professor Higgins made Eliza Doolittle go through to change her accent in *My Fair Lady*? This section tells you more about it.

Some researchers have argued that we have an automatic tendency to pay attention to the way others speak. When we talk to someone, we look at their vocabulary (*supper* vs. *dinner*), morphological variations (*going* vs. *goin*), dialectal variations (*cookie* vs. *biscuit*) and accent (*brother* vs. *broder*). We use this information to categorize people into different social groups, with the difference between one's own group and the other's group being the most important. In fact, it has been argued that this bias can be a bigger determinant of grouping people than other characteristics such as skin colour. This argument is based on the idea that our ancestors had few opportunities to interact with people who varied significantly in their physical characteristics (such as skin colour). However, they did probably have many more interactions with people who spoke another language, or at least who showed enough variation in their speech to allow them to be classified as belonging to their same group or not. Because of this, language use has been more relevant in evolution than other properties whose variability was less frequent in prehistoric times. And if you stop to think about it, you will see how much information we get from just a few words: whether the individuals are from the same country or not, their socio-economic status, and so on. The Bible talks about how the pronunciation of a single word, *Shibboleth*, served to differentiate between people from two tribes, since the first sound was pronounced distinctly (remember *perceptual adaptation* described in Chapter 1). Those (a few thousand) who did not pronounce the first sound correctly died beheaded by the other tribe.

This hypothesis was tested in an ingenious experiment conducted by David Pietraszewski at the University of California, Santa Barbara. Let me explain it in some detail. A series of photos with faces of people are presented to participants. Each time one of the faces appears, the participant hears a phrase. Each face is shown three times with different associated phrases. Of the eight faces, half say phrases in English with a British accent and the other half with an American accent. All of the participants had English as their mother tongue and were from the United States. In the exposure phase, the participants simply had to listen to the phrases and look at the faces. Nothing more. Once this step was finished, the eight faces were shown on the computer screen and the phrases from before were individually played. Now the participants were asked who had said each phrase. That is, the question was who had said what. The task was quite difficult because since many sentences had been played, the participants' memories were quite poor. But that's exactly what the point was: that the participants get too confused to observe what kind of mistakes they make. The question is: if a phrase was attributed to the wrong person, had it been spoken with the same accent or not? In principle, errors should be distributed randomly, right? Well, there was actually a certain regularity to the participants' mistakes: they tended to more often choose someone who spoke with the same accent as the right person. It is as if during the exhibition phase the participants had automatically categorized faces according to the accent with which they spoke. This phenomenon of confusion doesn't only happen between two different accents of the native language, but also when the faces speak different foreign languages or even when one is the native language and another is a foreign language. By the way, when, after the experiment, the participants were asked whether they were aware of this bias, they said no, and they even claimed that their answers could not have been affected by such categorization.

Maybe this result does not surprise you too much. After all, isn't it possible that the same thing can happen with any characteristic that differentiates individuals? Well, this is indeed what happens. If we change the experiment and show the participants black and white faces, the phenomenon of confusion is the same. The same happens when people wear different university shirts. So the answer is yes, any clue can help us to categorize people and group them in different sets. So what's the big deal? In fact not all of these clues carry the same weight when we categorize: some weigh much more than others. The same experiment can be done by presenting two clues and seeing which one biases (or confuses) participants' answers to a larger degree. Now that there are black and white faces, and American and British accents, which one will have more weight in the process of categorizing? The result is most interesting in that accent skews the responses more than skin colour. A bit like the children above who were choosing their friends. In other words, it would seem that individuals take language use as a more indicative sign and not so much the colour of a person's skin.

Social categorization is often part of the development of stereotypes. Once we group people under a common umbrella, it is difficult not to assign each individual the properties that we believe the group has, whether good or bad. In the context of the foreign language, there is a somewhat problematic stereotype, especially for those of us who speak other languages in which we may have a foreign accent. It turns out that we tend to have more doubts about the truth of the facts described by speakers with a foreign accent compared to native speakers. For example, if we are asked to judge whether we believe the content is real in the phrase *Ants do not sleep*, we believe it to be more true when someone with a native accent says it rather than someone with a foreign accent. What's worse is that when we tell the participants that the speakers are simply reading phrases that the researcher has given them and that therefore the accuracy of the phrases has nothing to do with the person who reads them, the foreign-language effect is still present. That is, we believe someone who sounds like a native speaker more than someone who speaks with an accent.

It seems that when we interact with a person who has a foreign accent, we tend to process language somewhat differently than with native people. In some ways, and maybe due to certain problems with understanding, we pay less attention to the details of speech and look more at the communicative intent. It's a bit like we do not care what the person *says*, but what they really *mean*. And that is why our memory of the exact words that people use in a conversation is much more accurate when people speak with a native accent.

So when you speak in a foreign language, don't expect people to remember exactly what you said or the details of your message.

All of this shows just how strong language is as a factor of social categorization. Being conscious of it and understanding how these biases work is fundamental in reducing prejudices and unjustified discrimination of individuals and social groups.

And here we are at the end of our journey through the new science of bilingualism. I hope that you have enjoyed reading about how two languages can coexist in the same brain. This area of study is constantly evolving as we continue to explore the cognitive effects of bilingualism – and what it can tell us about our understanding of language, thought, and emotions.

Further Reading

Alexakis, V. (2006). *Foreign Words*. Bloomington, IN: Autumn Hill.

Armon-Lotem, S., de Jong, J., and Meir, N. (eds) (2015). *Assessing Multilingual Children: Disentangling Bilingualism from Language Impairment*. Bristol: Multilingual Matters.

Baus, C., and Costa, A. (eds) (2016). 'Second-language processing' [Special Issue], *Language Learning*, 66 (S2).

Blakemore, S.-J., and Frith, U. (2005). *The Learning Brain: Lessons for Education*. Maldon, MA, and Oxford: Wiley-Blackwell.

Dewaele, J.-M. (2010). *Emotions in Multiple Languages*. London: Palgrave Macmillan.

Friederici, A. (2017). *Language in our Brain: The Origins of a Uniquely Human Capacity*. Boston, MA: MIT Press.

Gazzaniga, M. (2006). *The Ethical Brain: The Science of our Moral Dilemmas*. New York, NY: Harper Perennial.

Grant, A., Dennis, N., and Li, P. (2014). 'Cognitive control, cognitive reserve, and memory in the aging bilingual brain', *Frontiers in Psychology*, 5, pp. 1–10.

Grosjean, F., and Li, P. (eds) (2012). *The Psycholinguistics of Bilingualism*. Oxford: Wiley-Blackwell.

Guasti, M. (2004). *Language Acquisition: The Growth of Grammar*. Cambridge, MA: MIT Press.

Gullberg, M., and Indefrey, P. (eds) (2006). *The Cognitive Neuroscience of Second Language*. Hoboken, NJ: Wiley.

Harris, S. (2011). *The Moral Landscape: How Science can Determine Human Values*. New York, NY: Free Press.

Hernandez, A. (2013). *The Bilingual Brain*. Oxford: Oxford University Press.

Hogarth, R. (2001). *Educating Intuition*. Chicago, IL: University of Chicago Press.

Kahneman, D. (2011). *Think Fast, Think Slow*. New York, NY: Farrar, Straus, and Giroux.

Karmiloff, K., and Karmiloff-Smith, A. (2001). *Pathways to Language*. Cambridge, MA: Harvard University Press.

Kemmerer, D. (2015). *Cognitive Neuroscience of Language*. London: Psychology Press.

Ledoux, J. (1998). *The Emotional Brain: The Mysterious Underpinnings of Emotional Life*. New York, NY: Simon and Schuster.

Pavlenko, A. (2014). *The Bilingual Bind and what it Tells us about Language and Thought*. Cambridge: Cambridge University Press.

Pinker, S. (2007). *The Stuff of Thought: Language as a Window into Human Nature*. New York, NY: Penguin.

Schwieter, J. W. (ed.) (2015). *The Cambridge Handbook of Bilingual Processing*. Cambridge: Cambridge University Press.

Tucker, A., and Stern, Y. (2014). 'Cognitive reserve and the aging brain', in A. Nair and M. Sabbagh (eds), *Geriatric Neurology*. Chichester: Wiley.

Image Credits

PLATES

1 R. Sebastian, A. Laird and S. Kiran, 'Meta-analysis of the neural representation of first language and second language', *Applied Psycholinguistics*, vol. 32, 4, 2011, pp. 799–819. © 2011, Cambridge University Press. All rights reserved.

2 M. Burgaleta, A. Sanjuán, N. Ventura-Campos, N. Sebastián-Gallés and C. Ávila, 'Bilingualism at the core of the brain. Structural differences between bilinguals and monolinguals revealed by subcortical shape analysis', *NeuroImage*, vol. 125, 2016, pp. 437–445. © 2015, Elsevier Inc. All rights reserved.

3 Jubin Abutalebi, Pasquale Anthony Della Rosa, David W. Green, Mireia Hernández, Paola Scifo, Roland Keim, Stefano F. Cappa and Albert Costa, 'Bilingualism Tunes the Anterior Cingulate Cortex for Conflict Monitoring', *Cereb Cortex*, 22 (9), 2012, pp. 2076–2086. Doi: 10.1093/cercor/bhr287. © 2011, Oxford University Press.

4 Gigi Luk, Ellen Bialystok, Fergus I. M. Craik and Cheryl L. Grady, 'Lifelong Bilingualism Maintains White Matter Integrity in Older Adults', *Journal of Neuroscience*, 31 (46), 2011, pp. 16808–16813. © 2017, Society for Neuroscience.

Index

Note: References in *italics* are to figures.